Atmospheric Pressure Plasma - from Diagnostics to Applications

Edited by Anton Nikiforov and Zhiqiang Chen

Published in London, United Kingdom

IntechOpen

Supporting open minds since 2005

Atmospheric Pressure Plasma - from Diagnostics to Applications
http://dx.doi.org/10.5772/intechopen.75279
Edited by Anton Nikiforov and Zhiqiang Chen

Contributors
Qing Xiong, Jin Li, Boxue Du, Hucheng Liang, Floran Peeters, Tom Butterworth, Felipe Miranda, Alexei Essiptchouk, Gilberto Petraconi, Felipe Caliari, Yury Gorbanev, Annemie Bogaerts, Kazuo Shimizu, Marius Blajan, Jaroslav Kristof

First published in London, United Kingdom, 2019 by IntechOpen
IntechOpen is the global imprint of INTECHOPEN LIMITED, registered in England and Wales, registration number: 11086078, The Shard, 25th floor, 32 London Bridge Street
London, SE19SG - United Kingdom
Printed in Croatia

British Library Cataloguing-in-Publication Data
A catalogue record for this book is available from the British Library

Additional hard copies can be obtained from orders@intechopen.com

Atmospheric Pressure Plasma - from Diagnostics to Applications
Edited by Anton Nikiforov and Zhiqiang Chen
p. cm.
Print ISBN 978-1-83880-249-3
Online ISBN 978-1-83880-250-9

For EU product safety concerns:
IN TECH d.o.o., Prolaz Marije Krucifikse Kozulić 3, 51000 Rijeka, Croatia, info@intechopen.com or visit our website at intechopen.com.

Meet the editors

Anton Nikiforov obtained his PhD degree in 2004 at the Russian Academy of Sciences. From 2009, Dr. Nikiforov has been working at Ghent University in the Department of Applied Physics. His main research activity is experimental work that focuses on the physics, engineering, and applications of nonequilibrium ("cold") atmospheric pressure plasmas. Ongoing physical activities include the study of plasma-surface interactions, plasmas in liquids, and advanced diagnostics in nonequilibrium discharges. The engineering part of the research involves the development and construction of novel types of atmospheric pressure plasma sources and plasma technologies.

Dr. Zhiqiang Chen received his PhD degree in Plasma Surface Engineering from Deakin University in 2013. He is currently working at Deakin Plasma Research Group on the development of smart plasma coatings for industrial demands. In the past, he coordinated a number of industrial projects on plasma technology. His current research is focused on the coordination of projects for developing new large-scale plasma coating systems. His research activities cover the development of liquid and atmospheric plasma technologies and methodologies, and engineering materials with new functionalities, for applications ranging from energy, composites, waste gas, and water treatment, to nanofabrication and textiles.

Contents

Preface

Plasma as the fourth state of matter differs from solid, liquid, and gaseous states and is an ionized gas consisting of both negative and positive ions, electrons, neutral atoms, radicals, and ultraviolet photons. Plasma plays a crucial role in a wide variety of applications in material processing, environmental control, semiconductor manufacturing, biomedicine, green energy, and many more. Generally, plasma can be divided into two categories, namely low-pressure and atmospheric-pressure plasma. Compared to low-pressure plasma, atmospheric-pressure plasma overcomes the drawbacks associated with vacuum, which makes it the most sought-after plasma process in industrial applications. Over the last few decades, atmospheric-pressure plasmas have started to attract increasing attention from both scientists and industry due to a variety of potential applications. Because of increasing interest in this topic, the focus of this book is on providing engineers and scientists with a fundamental understanding of the physical and chemical properties of different atmospheric-pressure plasmas via plasma diagnostic techniques and their applications.

The book has been organized into two parts. Part I focuses on the latest achievements in advanced diagnostics of different atmospheric-pressure plasmas and contains three chapters. Chapter 1 introduces advanced optical diagnostics to visualize the physical and chemical properties of atmospheric-pressure plasmas. Optical diagnostic techniques include optical emission spectroscopy, Schlieren imaging or shadowgraph, and invasive methods of active laser spectroscopy, including laser-induced fluorescence, laser or broadband absorption, cavity ring-down spectroscopy, and laser scattering. Chapter 2 not only reviews electrical diagnostic techniques to measure charge-voltage signals of atmospheric-pressure dielectric barrier discharges (DBD) but also provides information at length on the interpretation of charge-voltage diagram results from different practical cases. It is a great reference for those who work on charge-voltage measurement techniques to understand or solve the problems they encounter. Chapter 3 is focused on the development of a model for charge accumulation on the surface of epoxy materials, which is an important problem in the development of insulators for high-voltage applications and electrical power transmission lines.

Part II from Chapter 4 to Chapter 6 focuses on the applications of different atmospheric-pressure plasmas and the measurement of plasma properties, which are of key importance for plasma applications, including the use of plasma in the biomedical field. Chapter 4 demonstrates the various applications of DBD microplasma from indoor air purification and skin treatment for drug delivery, to particle removal and the plasma actuator for flow control, while Chapter 5 reviews atmospheric/thermal-plasma spray processes and their applications in the formation of nanostructure coatings. Chapter 6 discusses diverse chemical approaches to detect reactive oxygen and nitrogen species induced in aqueous media by atmospheric-pressure plasma in regard to the biomedical application of plasmas.

We hope that with this book the readers will be introduced to the exciting world of plasma applications and plasma diagnostics, where numerous discoveries for both fundamental and applied science are foreseen for future studies.

Dr. Anton Nikiforov
Department of Applied Physics,
Ghent University,
Belgium

Dr. Zhiqiang Chen
Institute for Frontier Materials,
Deakin University,
Australia

Section 1

Plasma Diagnosctics

Chapter 1

Advanced Optical Diagnostics of Atmospheric Pressure Plasma

Qing Xiong

Abstract

Atmospheric-pressure plasma has been employed in various applications including bio-medicine, environmental pollution control, material processing. Diagnostic characterization of plasma sources is critical and indispensable for plasma control and achieving optimized treatment efficiency. In this chapter we will introduce several advanced optical techniques to visualize the detailed physical-and-chemical properties of atmospheric-pressure discharges. Non-invasive approaches of optical emission spectroscopy (OES), schlieren or shadowgraph, invasive methods of active laser spectroscopy including laser-induced fluorescence (LIF), laser or broadband absorption, cavity ring-down spectroscopy (CRDS), and laser scattering are illustrated. Basic plasma parameters of gas temperature, electron density and temperature, electric field strength, and reactive chemical gaseous species (O, H, N, OH, NO, O_3, etc.) are able to be monitored. Comparisons and comments of these approaches are provided depending on diagnostic purposes.

Keywords: advanced optical techniques, plasma parameters, active laser spectroscopy, LIF, absorption spectroscopy

1. Introduction

Atmospheric pressure plasmas attract intensely interests because of non-need of costly vacuum requirement and enabling in-line processing. Prospective novel applications have emerged over recent years, e.g., plasma bio-medicine, material synthesis, pollutant degradation, chemical conversion. However, with increase of gas pressure, plasmas tend to become constricted, unstable, and non-uniform. Miniaturization of discharge volume with non-equilibrium kinetic property are typically characterized by atmospheric pressure plasmas. And in many cases they are of strong gradients in space and in time as well, e.g., the so-called microplasmas. Much higher diagnostic requirements are put forward to atmospheric pressure plasmas with high time-and-space resolution. Not only the traditional optical emission spectroscopy (OES) accessing emitting species, but also advanced active diagnostics particularly the laser-aided techniques probing plasma parameters (T_g, n_e, T_e) and reactive plasma species, are the topics discussed in this chapter.

2. Diagnostics of basic plasma parameters

Plasma parameters mainly are the gas temperature T_g, electron density n_e and temperature T_e. The three are important physical factors directly influencing the plasma generation and physical-and-chemical properties. Diagnostics of the three parameters are necessary in plasma field from points of both fundamental research and applications.

2.1 Gas temperature

From points of both fundamental investigation and applications, gas temperature (T_g) is one of the important basic plasma parameters strongly affecting the discharge property, especially the plasma chemistry as most chemical reactions are temperature dependent, and thus influences the final outcomes of plasma applications. Standard direct approach by thermocouples is difficult due to induced strong effects on plasma properties. Although thermometric probes based on fiber-optics are developed and applied to plasma diagnostics [1], the use of this technique is still rare and limited to, e.g., microplasmas.

Rotational distributions of molecules, typically from excited molecules are often used to estimate the plasma gas temperature, if the former achieves an equilibrium Boltzmann distribution [2, 3]. In this case the distribution of interested rotational manifold can be estimated from corresponding emissions in UV to visible range and easy to be detected by a spectrometer. Excited species $OH(A^2\Sigma^+)$ and $N_2(C^3\Pi_u)$ are often used for the purpose of T_g estimation, by using programs of Lifbase or SpecAir [4, 5]. However, it needs to keep in mind that, on one hand this method is strongly affected by the production process of excited molecular state. Its rotational population distribution may highly deviate from thermal equilibrium state. This is typically characterized by atmospheric pressure plasmas, in which fast collisions significantly reduce the lifetime of rotational states and if the rotational energy transfer (RET) process is not fast enough to thermalized all rotational levels, a non-thermalized distribution is resulted. In this case T_g estimation will be not precise and overestimated due to over-populated high rotational states. This situation is often existent in water-contacting discharges if $OH(A^2\Sigma^+)$, or argon discharges if $N_2(C^3\Pi_u)$, are applied for T_g determination [6–9]. In this case, method of Boltzmann plotting may help by using the part of rotational states with small rotational number, as these levels with low rotational energy potential are easily thermalized by RET process [2].

On the other hand, it should mention that the emitting region usually characterizes only a part but not the whole discharge space. Commonly, it corresponds to the hot core of the discharge [10]. Other advanced approaches are desired if the full-space distribution of T_g is needed, for instance to estimate the heat flux of discharge to treated object. Laser scattering techniques such as Rayleigh scattering and Raman scattering, and calibrated schlieren photography are proposed and will be illustrated in this chapter. All these methods depend on the neutral density which directly relates to T_g if the ideal gas principle is valid.

Rayleigh scattering is elastic scattering on heavy species and its signal has the same wavelength as the laser. Its intensity depends proportionally on the scattering cross section, laser intensity, and amount of neutral particles probed in the scatter volume [11–14]. By comparing the scattering signals of plasma off I^R_{off} and plasma on I^R_{on}, the gas temperature is obtained if the gas composition does not vary significantly in the probed volume:

$$T_g = T_{ref}\frac{I^R_{off}}{I^R_{on}} \tag{1}$$

where T_{ref} is the known room temperature when plasma is off. An example of a Rayleigh scattering measurement in a pin-to-pin micro-glow air discharge is shown in **Figure 1**. This technique is favored for capturing the T_g spatial distribution of atmospheric pressure plasmas directly, but still following issues needs to be considered if it is applied:

- Laser-induced breakdown needs to be avoided by using appropriate laser energy. UV laser wavelength is suggested for this purpose, even laser energy is low it still able to produce intense scattering signal as the cross section has a strong wavelength dependence ($\sigma \propto \lambda^{-4}$) [12].

- Mie scattering on dust particles and stray scattering light from electrodes needs to be avoided. Therefore, this technique is quite difficult for probing discharge regions close to electrodes.

- The scattering signal will be weak for helium discharge due to the low Rayleigh cross section of helium atoms.

- Deviation will increase for temperature level above 2000 K as intensity of I_{on}^{R} will be very low in above equation.

Schlieren photography also is applied to measure the gas density gradients and correspondingly, to estimate the gas temperature [15, 16]. The advantage of this approach is achievable of the whole T_g distribution by only one schlieren image, if after calibration. A schlieren image visualizes the deflection of light rays along the optical path. The signal intensity of each pixel on the schlieren image is an integration result of bent light ray which strongly depends on the density distribution of gas species. If the gas composition is known and its spatial gradient is induced by thermal motion, then T_g map is able to be calculated through Abel inverse integral transformation. Symmetric assumption is usually made for the T_g map in non-uniform discharges such as pin-to-pin microplasma. For asymmetric discharges, it

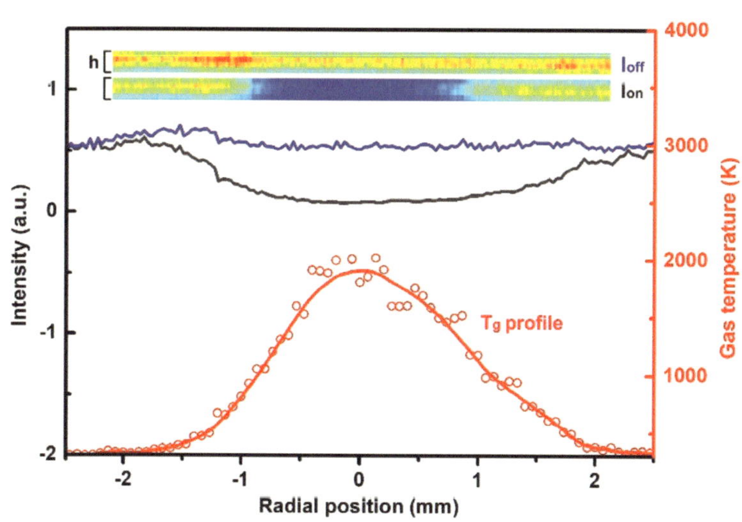

Figure 1.
Radial profiles of Rayleigh scattering intensities when the discharge was off and on, and the estimated gas temperature profile in the middle of discharge column with current of 8 mA [10].

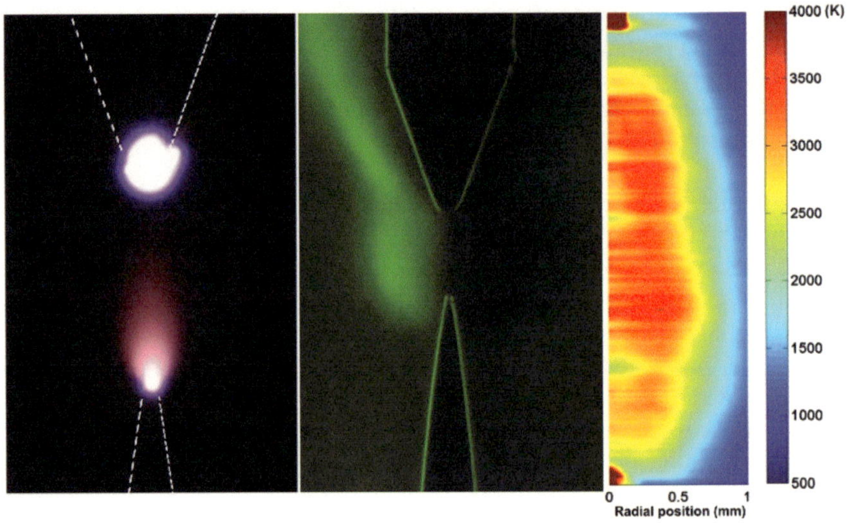

Figure 2.
(From left to right) A pin-to-pin air micro-glow discharge under 16 mA current, its Schlieren image, and two-dimensional T_g map.

is difficult but still possible by a computed tomographic reconstruction technique and multi-directional calibrated Schlieren imaging [16, 17].

Figure 2 shows an example of full T_g map estimated by calibrated schlieren photography [10]. It needs to point out that this technique suffers the optical diffraction effect on the object boundary, which usually induces an overestimation of temperature. However, compared to laser scattering approach, calibrated schlieren is a low-cost way for full-mapping gas temperature in discharges. Its drawback is the calibrated calculation process and need extra calibration by a long-focus lens.

Other methods including Raman scattering also allows one to measure the gas temperature [13]. It is the inelastic scattering of light on molecules such as N_2 or O_2 and its signal depends on neutral density, similar to the Rayleigh scattering. However, this technique requires expensive laser equipment and filter system for removing strong interference of Rayleigh scattering. Doppler broadening and Van der Waals broadening are analyzed for line profiles to estimate T_g as well if the Stark broadening is not dominant [18, 19]. However, the broadening routines require high spectral resolution to reduce the effect from instrumental broadening. Compared to emission lines, fine structure of atomic absorption line is achievable by tunable diode laser absorption measurement [8]. Compared to Doppler broadening, Van der Waals broadening is preferred for atmospheric-pressure plasmas with low gas temperature as the latter depends strongly on the neutral density. It should mention that the both broadening routines are line-integrated measurements and Abel inversion is required if radial profile of T_g is desired. **Table 1** concludes and compares all these methods for T_g diagnostics described above. Two or more techniques are suggested if diagnosing non-uniform high-pressure plasmas.

2.2 Electric-field strength

Electric-field strength E determination is always a tough issue for high-pressure plasmas. Compared to emission spectroscopic methods based on Stark effect of atomic lines, coherent Raman scattering (CRS) technique is favored for both static and fast transient discharges at elevated pressures [22, 23]. Its basic principle is vibrational excitation of Raman-active molecular which produces an infrared (IR)

Methods	Advantages	Disadvantages	Ref.
Rotational temperature of molecules	• Spatial resolution can be high; • Flexible and achievable by a fiber-optics spectrometer	• Line-averaged • Non-equilibrium distribution if quenching is strong • Characterize only radiative region	[2, 3]
Broadenings of line profile	• Spatial resolution can be high; • Non-dependent on quenching process	• Require high-spectral resolution; • Line-averaged	[18, 19]
Rayleigh scattering	• Spatial resolution can be high; • 2D measurement; • Accuracy is high for $T_g < 2000$ K	• Require expensive laser system; • Stray light effect; • Deviation increase for $T_g > 2000$ K	[12–14]
Raman scattering	• Spatial resolution can be high; • 2D measurement	• Require expensive laser system and notch filter	[13, 14]
Schlieren imaging	• Spatial resolution can be high; • Full spatial-field detection	• Need calibration and complex calculation	[10, 15–17, 20, 21]

Table 1.
Summary comparison of various diagnostic techniques for gas temperature determination in atmospheric-pressure discharges.

coherent beam under the effects of two extra nanosecond pulsed laser beams and an electric field. The IR light can be considered as an anti-Stokes wave produced by the electric fields of two laser beams and a third field at zero frequency, i.e., the probed dc electric field [22, 24, 25]. At the same time the laser beams also produce conventional coherent anti-Stokes Raman scattering (CARS) regardless of the presence of the dc electric field. And E is able to be determined from the detected intensities of the two coherent Raman scattering beam, according to:

$$\mathrm{I}_{IR} = \mathrm{C}_1\left(N_g - N_{ex}\right)^2 I_1 I_2 E^2 \tag{2}$$

$$\mathrm{I}_{CARS} = \mathrm{C}_2\left(N_g - N_{ex}\right)^2 I_1^2 I_2 \tag{3}$$

$$\mathrm{E}^2 = \frac{\mathrm{C}_2}{\mathrm{C}_1}\frac{\mathrm{I}_1\mathrm{I}_{\mathrm{IR}}}{\mathrm{I}_{\mathrm{CARS}}} = \mathrm{C}_0\frac{\mathrm{I}_{\mathrm{IR}}}{\mathrm{I}_{\mathrm{CARS}}} \tag{4}$$

N_g and N_{ex} are the number densities of scattering molecules at the ground and excited levels involved in the vibrational transition. Since the incident laser intensity I_1 is known and can be treated as a constant with two other constants C_1 and C_2. The value of parameter C_0 is almost independent of the gas density or temperature which significantly simplifies the analysis of this technique for E measurement in discharges.

Molecules including N_2 and H_2 have been used as the probe species, particularly the latter due to its large Raman cross section [22, 25]. However, compared to H_2 nitrogen gas has significant potentials for diagnosing atmospheric-pressure discharges especially discharges mixing with nitrogen gas. Based on the four-wave mixing principle of CRS technique by two laser beams, recently a novel and simpler CRS scheme was developed by *I. Adamovich*' group in Ohio State University by using only one laser beam [26–29]. A high-pressure gas cell filled with N_2 and He with 1:1 mixture ratio was used and pumped by the same laser beam to produce the 607 nm first order Stokes signal of N_2. The 607 nm Stokes beam was generated to replace the need of another nanosecond laser system used in traditional CRS technique. Then the pump and Stokes beams are focused in the discharge region to produce the IR anti-Stokes beam under the presence of probed dc electric field [28]. They applied a picosecond laser at 532 nm especially for probing the transient electric field in nanosecond discharges.

A further simpler approach based on the second harmonic generation (SHG) was developed by the same group for diagnosing the time-resolved evolution of the electric field vector in a nanosecond atmospheric-pressure surface plasma [29]. A single picosecond laser was used without Stokes gas cell. The pump laser wavelength is 1064 nm and focused to the discharge region and under the presence of an external electric field, SHG at 532 nm is generated with intensity proportional to the probed E of discharge [29]. Compared to CRS measurements, this SHG technique involves only one nonlinear optical process which enables better shot-to-shot signal reproducibility and signal/noise level. By using a polarizer in front of the SHG detector the vertical and horizontal components of E in discharge are able to be measured separately. Absolute calibration is required for above all approaches by measuring a known Laplacian electric field between two parallel metal plate electrodes. It should be mentioned that low-frequency nanosecond lasers, such as 10 Hz Nd:YAG pumping laser, is also suitable for E measurement in atmospheric-pressure discharges if ns time-resolution is not required.

2.3 Electron density and temperature

Electron density n_e and temperature T_e are two important fundamental plasma parameters since plasma property is mostly drove by electrons. Stark broadening of atomic line profile is often applied for estimating electron density if n_e typically exceeds 10^{14} cm^{-3} in atmospheric-pressure plasmas. The H_β line is most frequently used for the diagnostic purpose of n_e due to its strong Stark broadening but less self-absorption and weak effect from ion dynamics in discharges [30, 31]. Other hydrogen Balmer lines including H_α and H_γ are also studied and used for n_e determination. The H_α line is more intense of emission but easily suffered self-absorption effects. The Stark broadening of hydrogen Balmer lines has been studied extensively in [32] by group of M. Gigosos. They calculated the line widths of H_β Stark broadening for wide ranges of electron density, temperature and emitter-perturber reduced mass. An equation is derived for n_e based on the full-width at half-maximum (FWHM) of H_β Stark broadening profile [32].

$$w_s = 4.8\ \text{nm} \left(\frac{n_e}{10^{17}\text{cm}^{-3}}\right)^{0.68116} \tag{5}$$

The Stark broadening of H_β line facilitates the n_e determination in atmospheric-pressure plasmas. High-spectrum resolution (e.g., 10 pm) is recommended for capturing a clear shape of the H_β line profile since other broadenings including Doppler, van der Waals, and instrumental broadenings contribute also to the detected line profile. The effects of these broadenings need to be separated to obtain the Stark part and accurate deconvolution procedures normally are required for this purpose [33]. Other atomic lines are also possible for n_e measurement based on their Stark broadening if hydrogen gas is absent, for example, He (447.1 nm) or Ar (415.9 nm) lines [34, 35].

Microwave interferometry and CO_2-laser heterodyne interferometry have been applied as well to n_e measurement in plasmas based on the principle of change in refractive index of electromagnetic radiation passing through the discharge region [36, 37]. The main advantage of this technique is that it is able to measure electron densities at order of 10^{13} cm^{-3} lower than that by the Hβ Stark line broadening. However, since it is a line-integrated method and thermal effect may contribute also to the detected phase shift of probed electromagnetic radiation. This issue can be solved by using two laser sources with different wavelength [36]. Abel inversion is necessary to obtain spatially resolved electron density.

Compared to above two line-integrated techniques, Thomson scattering is a more advanced optical approach allows one probe directly information of both n_e and T_e [13]. This technique is based on the elastic scattering of light on free electrons, and its signal is Doppler broadened due to the velocity of electrons if the electron energy distribution function (EEDF) is a Maxwell-Boltzmann distribution. Therefore, n_e is able to be determined from the measured scattered power per unit of wavelength S_λ [38]:

$$S_\lambda = Cn_e \frac{d\sigma}{d\Omega} G_\lambda(\lambda) \tag{6}$$

where $C = flI_l \Delta\Omega$ is a constant factor depending on the optics efficiency f, length of detection volume l, laser power I_l, and detection solid angle $\Delta\Omega$. Absolute calibration needs to be done to determine the C value normally by means of Raman scattering on ambient air at atmospheric-pressure and room temperature [13]. $d\sigma/d\Omega$ is the differential cross section and $G_\lambda(\lambda)$ is the function of spectral distribution and normalized normally that $\int G_\lambda(\lambda)d\lambda = 1$. T_e can be obtained by the measured Gaussian profile of Thomson scattering signal according to [13, 39]:

$$T_e = \frac{m_e c^2}{4k_B}\left(\frac{\Delta\lambda}{\lambda_l}\right)^2 \tag{7}$$

$\Delta\lambda$ is the $1/e$ width of the Gaussian profile. Although these advantages of this advanced diagnostic technique, its implementation is not straightforward in general due to the weak scattered signal typically orders of magnitude lower than that of Rayleigh scattering in addition to stray light. A notch filter or a triple grating spectrometer, or an atomic absorption cell, are necessary to capture the weak scattered light [40–42]. And, laser-induced optical effects such as photo-detachment, photon-ionization, or electron heating may occur if high laser power is used in measurements, and need to be checked by measuring the signal as a function of laser power. More details about the Thomson scattering approach can be found in [13, 39, 41, 43–48].

3. Diagnostics of plasma chemistry (reactive species)

Plasma chemistry covers broad ranges of chemical properties of discharge plasmas, and rapidly grows to an important research area of scientific endeavor holding great promise for many novel plasma applications. Reactive species, including neutrals, metastable or excited species, ions consist the dominant active agents of discharges and play critical roles in plasma treatments, e.g., in fields of medicine, material, energy, etc. Various diagnostic techniques have been developed and applied for probing different reactive species quantitatively. And in this chapter three important techniques based on principles of laser-induced fluorescence (LIF) and absorption spectroscopy, are included.

3.1 Laser-induced fluorescence spectroscopy

Laser-induced fluorescence (LIF) is an *in situ* sensitive technique monitoring transient species with high temporal and spatial resolution [49]. This approach is based on a two-step process, where atoms or molecules in the initial electronic state i are firstly pumped to an upper electronic state j by absorbing laser photons with energy $\Delta E = h\upsilon_{ij}$ same to the transition gap between states i and j. Secondly, the excited state j undergoes spontaneous transition to a lower state k and emits photons with energy $h\upsilon_{jk}$ called the fluorescence signal. As an example, a conceptual LIF scheme of OH radicals, one typical reactive molecular species studied frequently by this technique, is depicted in **Figure 3**. For molecules fast vibrational energy transfer (VET) and rotational energy transfer (RET) processes are occurring after laser excitation, as well as the pre-dissociation process.

For atomic species these ro-vibrational processes are not existent. Although LIF is a direct technique allows us detect the spatial distributions of interested species, absolute calibration and analysis model are required for quantitative measurements [50–54]. For the case of single-photon excitation the fluorescence calibration is relative simple with the aid of Rayleigh scattering. It is based on a comparison under identical conditions of laser irradiation in Rayleigh scattering and fluorescence detection in LIF. A reference gas with known density is usually used for Rayleigh calibration, for instance air gas under room temperature. Typical laser excitations and fluorescence transitions of various species are listed in **Table 2** together with suggested calibration.

Analysis of fluorescence signal for absolutely determining probed species density can be very complicated under high gas pressure due to the fast collisional processes [67]. In this case the excited state after laser pumping may redistribute because of fast collisions at same timescale or even faster and these processes have to be considered in the analysis model. This could be more complicated for molecular species, for instance OH, NO, etc. in addition to the fast processes of VET and

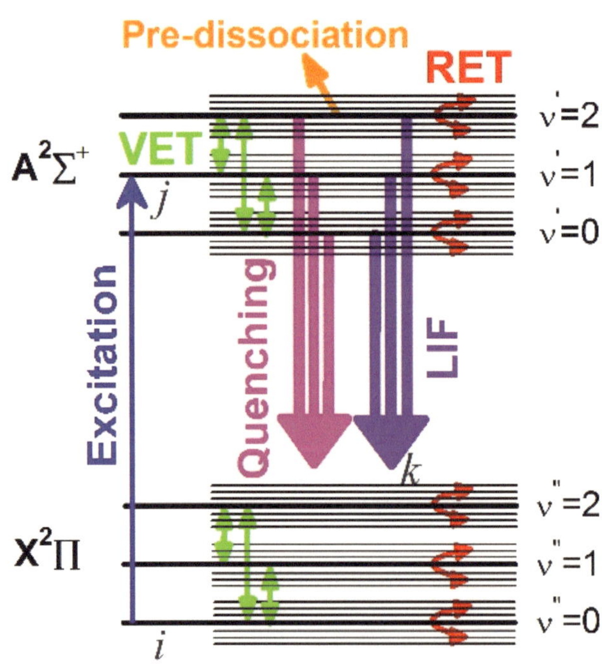

Figure 3.
LIF scheme of OH radicals.

Species	Excitation transition	Fluorescence transition	Calibration	Ref.
OH	$X^2\Pi$, v = 0 → $A^2\Sigma^+$, v = 1 283 nm	$A^2\Sigma^+$, v = 0 → $X^2\Pi$, v = 0 308 nm	Rayleigh of air	[50–54]
NO	$X^2\Pi$, v = 0 → $A^2\Sigma^+$, v = 0 226 nm	$A^2\Sigma^+$, v = 2 → $X^2\Pi$, v = 0 247 nm	NO gas	[55–59]
$N_2(A^3\Sigma_u^+)$	$A^3\Sigma_u^+$, v = 0 → $B^3\Pi_g$, v = 4 618 nm	$B^3\Pi_g$, v = 4 → $A^3\Sigma_u^+$, v = 1 678 nm	Rayleigh	[60–62]
NH	$X^3\Sigma^-$, v = 0 → $A^3\Pi_i$, v = 0 334 nm	$A^3\Pi_i$, v = 0 → $X^3\Sigma^-$, v = 0 336 nm	Rayleigh	[63, 64]
CH	$X^2\Pi$, v = 0 → $A^2\Delta$, v = 1 364 nm	$A^2\Delta$, v = 0 → $X^2\Pi$, v = 0 405 nm	Rayleigh	[65, 66]

Table 2.
The plasma species have been measured by single-photon LIF.

Species	Excitation wavelength (nm)	Fluorescence wavelength (nm)	Calibration	Ref.
O	225	845	Xe gas	[49, 73]
H	205	656	Kr gas	[74]
N	206	742–747	Kr gas	[75]
F	170	420	Ar gas	[49]
C	246	477	Rayleigh	[49]

Table 3.
Excitation schemes of various atomic species by TA-LIF technique.

RET [49]. Assumptions are often made for simplifying the LIF model, e.g., RET is fast enough to redistribute the rotational manifold of excited vibrational state and in this case the lifetime of the vibrational band can be considered only depending on quenching and emission coefficients. Effective lifetime of excited vibrational band instead of single rotational level, is used in the LIF model. And this is normally correct if RET process is fast, but still need to check if concentrations of quenching species such as H_2O to OH is very high [54, 68–70]. Furthermore, the laser power needs to be kept low and linear LIF is achieved to avoid stimulated emission and large depletion of initial pumped state density.

In many cases the excitation radiations locate range of deep UV (e.g., <100 nm) and are strongly absorbed by ambient gaseous medium (e.g., air or water molecules), then two-photon excitation in the UV spectral range (>200 nm) can be applied for the same diagnostic purpose, namely the so-called two-photon absorption LIF (TA-LIF) [49]. Compared to single-photon LIF, TA-LIF is considerably less efficient and the excitation rate scales with the square of laser power. Normally higher laser power is required to produce detectable fluorescence intensity, but laser stimulated effects including photon-ionization, induced breakdown, stimulated emission, etc. needs to be avoided [71, 72]. **Table 3** lists excitation schemes of TA-LIF for various active atomic species in plasmas. An example of two-dimensional distribution of atomic oxygen in an atmospheric-pressure plasma jet-source by TA-LIF is presented in **Figure 4**.

3.2 Absorption spectroscopy

Another important diagnostic technique for atmospheric-pressure plasmas is absorption spectroscopy (AS) [77]. Compared to LIF or OES approaches, AS is an *in*

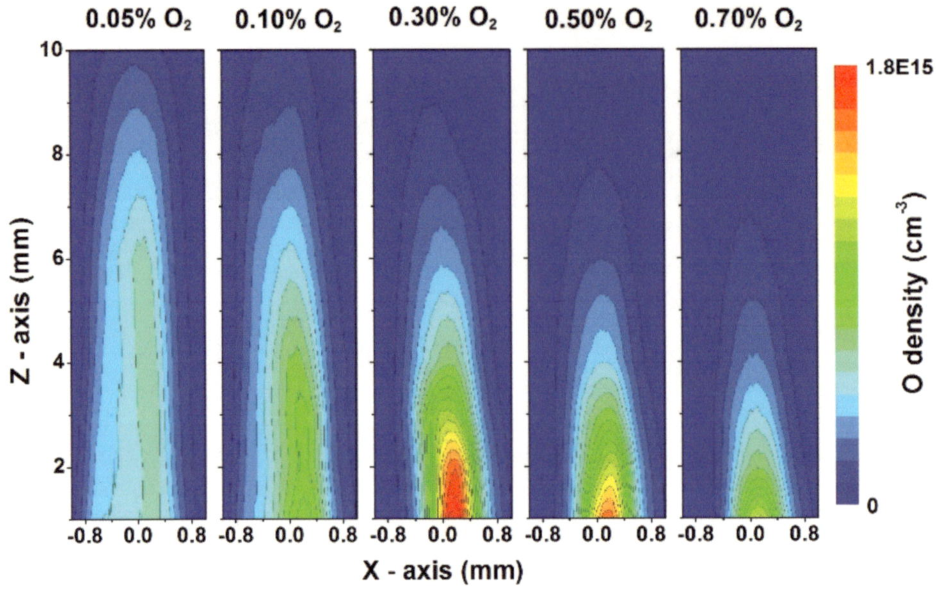

Figure 4.
Two-dimensional spatial distributions of ground-state O density in the x-z plane of the RF argon plasma jet under different O_2 admixtures [76].

situ and non-invasive technique probing absolute densities of interested species without need of extra calibration and independent on collisional quenching process. It relies on the *Beer-Lambert* law:

$$A(\lambda) = -\ln\left(\frac{I_s}{I_0}\right) = N\sigma(\lambda)L \tag{8}$$

where I_s is the light intensity passing through sample, I_0 is the incident radiation, N is species density, $\sigma(\lambda)$ is the wavelength dependent absorption cross-section, and L is the absorption path-length. If probed species density is low, the absorption length is normally increased by using a multi-pass spectroscopic absorption cell where light reflects between two mirrors, e.g., the *Herriott* cell [78]. However, if for micro-plasmas this technique maybe not sensitive enough if species density is not high to obtain a distinct absorbance signal.

The sensitivity of absorption measurements is strongly dependent on the stability of used light source. Various types of light sources, narrow-band or broadband with wavelength range from VUV to mid-infrared, are applied together with high spectral resolution detection system such as spectrometer or FTIR. AS by laser light sources, particularly tunable diode-lasers or quantum cascade lasers are attractive light sources have been applied to measure species densities in micro-plasmas [8, 79, 80]. These laser sources output specific wavelengths of large absorption cross-sections and enable fine scan of wavelength to obtain a clear absorption line profile, and from which species density and even gas temperature are able to be derived.

Wavelength selectivity is important in the implementation of AS technique as absorption may overlap by multi-species. And, it should be mentioned that absorption measurements yield line-integrated densities and spatial distribution is achievable by Abel inversion. Special efforts are required for non-uniform micro-plasmas by AS approach, since gas temperature and species density are spatially dependent. In this case broadband absorption can provide more details for deriving both the gas

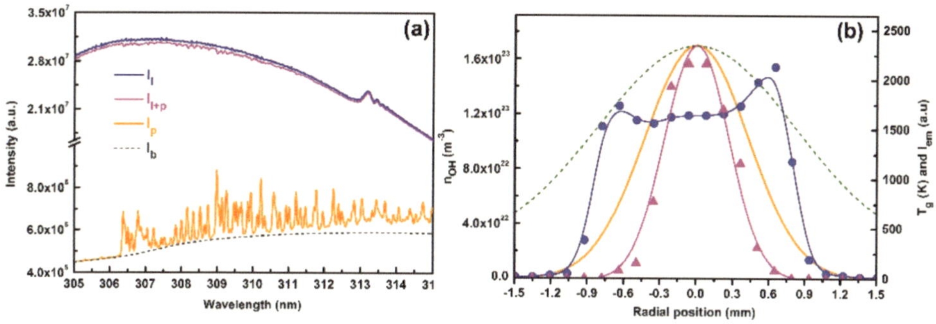

Figure 5.
(a) Measured spectra in the range 305–315 nm from the lamp source I_l, lamp and plasma on I_{l+p}, plasma on I_p, and the background I_b. (b) Radial profiles of OH density and gas temperature, and n(A-X) emission with or without filter from the cathode region of water-cathode discharge [83].

temperature and species density [81, 82]. An example of OH density determination by broadband UV absorption in an atmospheric-pressure pin-to-water discharge is depicted in **Figure 5** [83].

Another quite sensitive approach by using a cavity of two highly reflective mirrors is the cavity ring-down spectroscopy (CRDS) [84]. The laser pulse goes-and-back in the cavity and that effectively enhances the absorption path length in the probed gas sample by several orders of magnitude. The time-decay of light intensity leaking from the cavity depends on the reflectivity of mirrors and species density in the cavity. Based on that the species density is able to be obtained with the known mirror reflectivity. In addition to the high sensitivity, compared to absorption spectroscopy, CRDS is independent on fluctuations of light intensity since a time constant rather than an absolute absorbance is measured. More details of CRDS technique are included in Refs. [84-90].

4. Summary

High-end diagnostic techniques have been applied to atmospheric-pressure plasmas. Even for highly transient non-uniform micro-plasmas, reliable insights of discharge properties are able to be achieved. This chapter includes several important advanced diagnostic approaches specifically for plasma parameters and reactive plasma species. Two approaches are suggested for a same diagnostic purpose, for example, the determination of gas temperature or electron density, and compared to ensure and increase the measurement accuracy, particularly for atmospheric-pressure plasmas. Other important plasma properties, such as densities of ion species and ion flux, surface charges, have been investigated as well and related diagnostic techniques can be found in above (review) references.

Author details

Qing Xiong
State Key Laboratory of Power Transmission Equipment and System Security and New Technology, Chongqing University, Chongqing, China

*Address all correspondence to: qingugent@gmail.com

References

[1] Wertheimer MR, Ahlawat M, Saoudi B, Kashyap R. Accurate in-situ gas temperature measurements in dielectric barrier discharges at atmospheric pressure. Applied Physics Letters. 2012; **100**(20):201112

[2] Bruggeman P, Sadeghi N, Schram D, Linss V. Gas temperature determination from rotational lines in non-equilibrium plasmas: A review. Plasma Sources Science and Technology. 2014;**23**(2): 023001

[3] Bruggeman P, Verreycken T, Gonzalez MA, Walsh JL, Kong MG, Leys C, et al. Optical emission spectroscopy as a diagnostic for plasmas in liquids: Opportunities and pitfalls. Journal of Physics D: Applied Physics. 2010; **43**(12):124005

[4] Luque J, Crosley D. LIFBASE: Database and spectral simulation (version 1.9). SRI International Report MP. 1999. 99-009

[5] http://www.specair-radiation.net [database on the Internet]

[6] Bruggeman PJ, Kushner MJ, Locke BR, Gardeniers JGE, Graham WG, Graves DB, et al. Plasma–liquid interactions: A review and roadmap. Plasma Sources Science and Technology. 2016;**25**(5):053002

[7] Xiong Q, Nikiforov AY, Lu XP, Leys C. High-speed dispersed photographing of an open-air argon plasma plume by a grating–ICCD camera system. Journal of Physics D: Applied Physics. 2010; **43**(41):415201

[8] Bol'shakov AA, Cruden BA, Sharma SP. Determination of gas temperature and thermometric species in inductively coupled plasmas by emission and diode laser absorption. Plasma Sources Science and Technology. 2004;**13**(4): 691-700

[9] Jackson GP, King FL. Probing excitation/ionization processes in millisecond-pulsed glow discharges in argon through the addition of nitrogen. Spectrochimica Acta B. 2003;**58**(2): 185-209

[10] Xiong Q, Xu L, Wang X, Xiong L, Huang Q, Chen Q, et al. Full spatial-field visualization of gas temperature in an air micro-glow discharge by calibrated Schlieren photography. Journal of Physics D: Applied Physics. 2018;**51**(9):095207

[11] Miles RB, Lempert WR, Forkey JN. Laser Rayleigh scattering. Measurement Science and Technology. 2001;**12**(5): R33-R51

[12] Adams SF, Caplinger JE, Sommers BS. Spatial temperature mapping of an atmospheric microdischarge using ultraviolet Rayleigh scatter imaging. Plasma Sources Science and Technology. 2015;**24**(2):025031

[13] van Gessel AH, Carbone ED, Bruggeman P, van der Mullen JM. Laser scattering on an atmospheric pressure plasma jet: Disentangling Rayleigh. Raman and Thomson scattering. Plasma Sources Science and Technology. 2012; **21**(1):015003

[14] Verreycken T, van Gessel A, Pageau A, Bruggeman P. Validation of gas temperature measurements by OES in an atmospheric air glow discharge with water electrode using Rayleigh scattering. Plasma Sources Science and Technology. 2011;**20**(2):024002

[15] Hargather MJ, Settles GS. A comparison of three quantitative Schlieren techniques. Optics and Lasers in Engineering. 2012;**50**(1):8-17

[16] Settles GS, Hargather MJ. A review of recent developments in Schlieren and

shadowgraph techniques. Measurement Science and Technology. 2017;**28**(4): 042001

[17] Ishino Y, Hayashi N, Bt Abd Razak IF, Kato T, Kurimoto Y, Saiki Y. 3D-CT (computer tomography) measurement of an instantaneous density distribution of turbulent flames with a multi-directional quantitative Schlieren camera (reconstructions of high-speed premixed burner flames with different flow velocities). Flow, Turbulence and Combustion. 2016;**96**(3):819-835

[18] Ionascut-Nedelcescu A, Carlone C, Kogelschatz U, Gravelle DV, Boulos MI. Calculation of the gas temperature in a throughflow atmospheric pressure dielectric barrier discharge torch by spectral line shape analysis. Journal of Applied Physics. 2008;**103**(6):063305

[19] Yubero C, Dimitrijević MS, García MC, Calzada MD. Using the van der Waals broadening of the spectral atomic lines to measure the gas temperature of an argon microwave plasma at atmospheric pressure. Spectrochimica Acta B. 2007;**62**(2):169-176

[20] Qing X, Le X, Lin X, Qinghua H, Qiang C, Yi C, et al. Heat deposition in the thermal field of a micro-glow discharge: Effect of humidity. Plasma Sources Science and Technology. 2018; **27**(9):095010

[21] Tobin JD, Hargather MJ. Quantitative Schlieren measurement of explosively-driven shock wave density, temperature, and pressure profiles. Propellants, Explosives, Pyrotechnics. 2016;**41**(6):1050-1059

[22] Ito T, Kobayashi K, Mueller S, Luggenhölscher D, Czarnetzki U, Hamaguchi S. Electric field measurement in an atmospheric or higher pressure gas by coherent Raman scattering of nitrogen. Journal of Physics D: Applied Physics. 2009;**42**(9):092003

[23] Ochkin VN, Savinov SY, Tskhai SN, Czarnetzki U, Gathen VS, Dobele HF. Nonlinear optical techniques for plasma diagnostics. IEEE Transactions on Plasma Science. 1998;**26**(5):1502-1513

[24] Ito T, Kanazawa T, Hamaguchi S. Rapid breakdown mechanisms of open air nanosecond dielectric barrier discharges. Physical Review Letters. 2011;**107**(6):065002

[25] Ito T, Kobayashi K, Czarnetzki U, Hamaguchi S. Rapid formation of electric field profiles in repetitively pulsed high-voltage high-pressure nanosecond discharges. Journal of Physics D: Applied Physics. 2010;**43**(6): 062001

[26] Simeni MS, Baratte E, Zhang C, Frederickson K, Adamovich IV. Electric field measurements in nanosecond pulse discharges in air over liquid water surface. Plasma Sources Science and Technology. 2018;**27**(1):015011

[27] Simeni MS, Goldberg B, Gulko I, Frederickson K, Adamovich IV. Sub-nanosecond resolution electric field measurements during ns pulse breakdown in ambient air. Journal of Physics D: Applied Physics. 2017;**51**(1): 01LT01

[28] Simeni Simeni M, Goldberg BM, Zhang C, Frederickson K, Lempert WR, Adamovich IV. Electric field measurements in a nanosecond pulse discharge in atmospheric air. Journal of Physics D: Applied Physics. 2017;**50**(18): 184002

[29] Simeni Simeni M, Tang Y, Frederickson K, Adamovich IV. Electric field distribution in a surface plasma flow actuator powered by ns discharge pulse trains. Plasma Sources Science and Technology. 2018;**27**(10):104001

[30] Gigosos MA, Cardenoso V. Study of the effects of ion dynamics on Stark profiles of Balmer-α and -β lines using

simulation techniques. The Journal of Physics B: Atomic, Molecular and Optical Physics. 1987;**20**(22):6005-6019

[31] Ivković M, Jovićević S, Konjević N. Low electron density diagnostics: Development of optical emission spectroscopic techniques and some applications to microwave induced plasmas. Spectrochimica Acta B. 2004; **59**(5):591-605

[32] Gigosos MA, González MÁ, Cardeñoso V. Computer simulated Balmer-alpha, -beta and -gamma Stark line profiles for non-equilibrium plasmas diagnostics. Spectrochimica Acta B. 2003;**58**(8):1489-1504

[33] Bruggeman P, Schram D, González MÁ, Rego R, Kong MG, Leys C. Characterization of a direct dc-excited discharge in water by optical emission spectroscopy. Plasma Sources Science and Technology. 2009;**18**(2):025017

[34] Du B, Sadeghi N, Tsankov TV, Luggenhölscher D, Czarnetzki U. Temporally resolved optical emission spectroscopic investigations on a nanosecond self-pulsing micro-thin-cathode discharge. Plasma Sources Science and Technology. 2012;**21**(4): 045015

[35] Xiong Q, Nikiforov AY, González MÁ, Leys C, Lu XP. Characterization of an atmospheric helium plasma jet by relative and absolute optical emission spectroscopy. Plasma Sources Science and Technology. 2012;**22**(1):015011

[36] Choi J-Y, Takano N, Urabe K, Tachibana K. Measurement of electron density in atmospheric pressure small-scale plasmas using CO_2-laser heterodyne interferometry. Plasma Sources Science and Technology. 2009; **18**(3):035013

[37] Leipold F, Stark RH, El-Habachi A, Schoenbach KH. Electron density measurements in an atmospheric pressure air plasma by means of infrared heterodyne interferometry. Journal of Physics D: Applied Physics. 2000; **33**(18):2268-2273

[38] Kempkens H, Uhlenbusch J. Scattering diagnostics of low-temperature plasmas (Rayleigh scattering, Thomson scattering, CARS). Plasma Sources Science and Technology. 2000;**9**(4):492-506

[39] Muraoka K, Uchino K, Bowden MD. Diagnostics of low-density glow discharge plasmas using Thomson scattering. Plasma Physics and Controlled Fusion. 1998;**40**(7): 1221-1239

[40] Jia F, Sumi N, Ishikawa K, Kano H, Inui H, Kularatne J, et al. Laser scattering diagnosis of a 60-Hz non-equilibrium atmospheric pressure plasma jet. Applied Physics Express. 2011;**4**(2):026101

[41] Muraoka K, Kono A. Laser Thomson scattering for low-temperature plasmas. Journal of Physics D: Applied Physics. 2011;**44**(4):043001

[42] Sande MJ, Deckers RHM, Lepkojus F, Buscher W, Mullen JJAM. Time resolved electron density and temperature measurements on a capacitively coupled helium RF discharge. Plasma Sources Science and Technology. 2002;**11**(4):466-475

[43] Nedanovska E, Nersisyan G, Morgan TJ, Hüwel L, Lewis CLS, Riley D, et al. Comparison of the electron density measurements using Thomson scattering and emission spectroscopy for laser induced breakdown in one atmosphere of helium. Applied Physics Letters. 2011;**99**(26):261504

[44] Hassaballa S, Yakushiji M, Young-Kee K, Tomita K, Uchino K, Muraoka K. Two-dimensional structure of PDP micro-discharge plasmas obtained using laser Thomson scattering. IEEE

Transactions on Plasma Science. 2004; **32**(1):127-134

[45] Noguchi Y, Matsuoka A, Bowden MD, Uchino K, Muraoka K. Measurements of electron temperature and density of a micro-discharge plasma using laser Thomson scattering. Japanese Journal of Applied Physics. 2001;**40**(1(1)):326-329

[46] Kono A, Iwamoto K. High-spatial-resolution multichannel Thomson scattering measurements for atmospheric pressure microdischarge. Japanese Journal of Applied Physics. 2004;**43**(8A):L1010-L10L3

[47] Palomares JM, Iordanova E, Hübner S, Carbone EAD, JJAMvd M. Towards poly-diagnostics on cool atmospheric plasmas. Journal of Instrumentation. 2012;**7**(02):C02027-C

[48] Kempkens H, Uhlenbusch J. Scattering diagnostics of low-temperature plasmas (Rayleigh scattering, Thomson scattering, CARS). Plasma Sources Science and Technology. 2000;**9**:492

[49] Döbele H, Mosbach T, Niemi K, Gathen V. Laser-induced fluorescence measurements of absolute atomic densities: Concepts and limitations. Plasma Sources Science and Technology. 2005;**14**(2):S31-S41

[50] Dilecce G, Ambrico PF, Benedictis SD. An ambient air RF low-pressure pulsed discharge as an OH source for LIF calibration. Plasma Sources Science and Technology. 2004;**13**(2):237-244

[51] Dilecce G, Ambrico PF, Simek M, De Benedictis S. LIF diagnostics of hydroxyl radical in atmospheric pressure He-H_2O dielectric barrier discharges. Chemical Physics. 2012;**398**: 142-147

[52] Riès D, Dilecce G, Robert E, Ambrico P, Dozias S, Pouvesle J. LIF and fast imaging plasma jet characterization relevant for NTP biomedical applications. Journal of Physics D: Applied Physics. 2014;**47**(27):275401

[53] Verreycken T, van der Horst R, Baede A, Veldhuizen V, Bruggeman P. Time and spatially resolved LIF of OH in a plasma filament in atmospheric pressure He–H_2O. Journal of Physics D: Applied Physics. 2012;**45**(4):045205

[54] Yagi I, Ono R, Oda T, Takaki K. Two-dimensional LIF measurements of humidity and OH density resulting from evaporated water from a wet surface in plasma for medical use. Plasma Sources Science and Technology. 2015;**24**(1): 015002

[55] Iseni S, Zhang S, van Gessel AFH, Hofmann S, van Ham BTJ, Reuter S, et al. Nitric oxide density distributions in the effluent of an RF argon APPJ: Effect of gas flow rate and substrate. New Journal of Physics. 2014;**16**(12): 123011

[56] Burnette D, Shkurenkov I, Adamovich IV, Lempert WR. Kinetics of NO formation and decay in nanosecond pulse discharges in air, H_2-air, and C_2H_4-air mixtures. Plasma Sources Science and Technology. 2016; **25**(2):025012

[57] Dilecce G, Ambrico PF, Šimek M, De Benedictis S. New laser-induced fluorescence scheme for simultaneous OH and NO detection by a single laser set-up. Applied Physics B: Lasers and Optics. 2002;**75**(1):131-135

[58] van Gessel AFH, Hrycak B, Jasiński M, Mizeraczyk J, van der Mullen JJAM, Bruggeman PJ. Temperature and NO density measurements by LIF and OES on an atmospheric pressure plasma jet. Journal of Physics D: Applied Physics. 2013;**46**(9):095201

[59] van Gessel AFH, Bruggeman PJ. Thermalization of rotational states of

NO $A(2)\Sigma^+(v = 0)$ in an atmospheric pressure plasma. The Journal of Chemical Physics. 2013;**138**(20):204306

[60] Simek M, Ambrico P, Prukner V. LIF study of $N_2(A^3\Sigma_u^+, v = 0–10)$ vibrational kinetics under nitrogen streamer conditions. Journal of Physics D: Applied Physics. 2015;**48**(26):265202

[61] Simek M, Ambrico P, Prukner V. $N_2(A)$ vibrational kinetics in streamer discharges: Effect of oxygen on formation of low vibrational levels. APS Gaseous Electronics Conference 2016. Bochum 2016

[62] Ono R, Tobaru C, Teramoto Y, Oda T. Laser-induced fluorescence of $N_2(A^3\Sigma_u^+)$ metastable in N_2 pulsed positive corona discharge. Plasma Sources Science and Technology. 2009; **18**(2):025006

[63] Brackmann C, Zhou B, Li ZS, Aldén M. Strategies for quantitative planar laser-induced fluorescence of NH radicals in flames. Combustion Science and Technology. 2016;**188**(4–5):529-541

[64] Rensberger KJ, Dyer MJ, Wise ML, Copeland RA. Time-resolved laser-induced fluorescence in low-pressure flames. AIP Conference Proceedings. 1988;**172**(1):750-752

[65] Luque J, Klein-Douwel RJH, Jeffries JB, Smith GP, Crosley DR. Quantitative laser-induced fluorescence of CH in atmospheric pressure flames. Applied Physics B: Lasers and Optics. 2002; **75**(6):779-790

[66] Luque J, Crosley DR. Absolute CH concentrations in low-pressure flames measured with laser-induced fluorescence. Applied Physics B: Lasers and Optics. 1996;**63**(1):91-98

[67] Bruggeman P, Brandenburg R. Atmospheric pressure discharge filaments and microplasmas: Physics, chemistry and diagnostics. Journal of Physics D: Applied Physics. 2013; **46**(46):464001

[68] Yagi I, Shirakawa Y, Hirakata K, Akiyama T, Yonemori S, Mizuno K, et al. Measurement of OH, O, and NO densities and their correlations with mouse melanoma cell death rate treated by a nanosecond pulsed streamer discharge. Journal of Physics D: Applied Physics. 2015;**48**(42):424006

[69] Yonemori S, Nakagawa Y, Ono R, Oda T. Measurement of OH density and air–helium mixture ratio in an atmospheric-pressure helium plasma jet. Journal of Physics D: Applied Physics. 2012;**45**(22):225202

[70] Yonemori S, Ono R. Flux of OH and O radicals onto a surface by an atmospheric-pressure helium plasma jet measured by laser-induced fluorescence. Journal of Physics D: Applied Physics. 2014;**47**(12):125401

[71] Amorim J, Baravian G. The two-photon absorption laser induced stimulated emission as a hydrogen atoms diagnostic tool: Modelling and experiment. Optics Communications. 2001;**192**(3):277-286

[72] Amorim J, Baravian G, Touzeau M, Jolly J. Two-photon laser induced fluorescence and amplified spontaneous emission atom concentration measurements in O_2 and H_2 discharges. Journal of Applied Physics. 1994;**76**(3): 1487-1493

[73] Niemi K, Gathen VS-v, Döbele HF. Absolute atomic oxygen density measurements by two-photon absorption laser-induced fluorescence spectroscopy in an RF-excited atmospheric pressure plasma jet. Plasma Sources Science and Technology. 2005; **14**(2):375-386

[74] Yatom S, Luo Y, Xiong Q, Bruggeman PJ. Nanosecond pulsed humid Ar plasma jet in air: Shielding,

discharge characteristics and atomic hydrogen production. Journal of Physics D: Applied Physics. 2017;**50**(41):415204

[75] Wagenaars E, Gans T, O'Connell D, Niemi K. Two-photon absorption laser-induced fluorescence measurements of atomic nitrogen in a radio-frequency atmospheric-pressure plasma jet. Plasma Sources Science and Technology. 2012;**21**(4):042002

[76] Xiong Q, Liu H, Britun N, Nikiforov AY, Li L, Chen Q, et al. Time-selective TALIF spectroscopy of atomic oxygen applied to an atmospheric pressure argon plasma jet. IEEE Transactions on Plasma Science. 2016;**44**(11):2745-2753

[77] Reuter S, Sousa JS, Stancu GD, Hubertus van Helden JP. Review on VUV to MIR absorption spectroscopy of atmospheric pressure plasma jets. Plasma Sources Science and Technology. 2015;**24**(5):054001

[78] Reuter S, Winter J, Iseni S, Peters S, Schmidt-Bleker A, Dünnbier M, et al. Detection of ozone in a MHz argon plasma bullet jet. Plasma Sources Science and Technology. 2012;**21**(3): 034015

[79] Douat C, Hübner S, Engeln R, Benedikt J. Production of nitric/nitrous oxide by an atmospheric pressure plasma jet. Plasma Sources Science and Technology. 2016;**25**(2):025027

[80] Simeni Simeni M, Laux CO, Stancu GD. High-spatial resolution measurements of NO density and temperature by mid-IR QCLAS in open-air confined plasmas. Journal of Physics D: Applied Physics. 2017;**50**(27):274004

[81] Bruggeman P, Cunge G, Sadeghi N. Absolute OH density measurements by broadband UV absorption in diffuse atmospheric-pressure He–H_2O RF glow discharges. Plasma Sources Science and Technology. 2012;**21**(3):035019

[82] Dilecce G, Ambrico PF, Simek M, Benedictis SD. OH density measurement by time-resolved broad band absorption spectroscopy in an Ar–H_2O dielectric barrier discharge. Journal of Physics D: Applied Physics. 2012; **45**(12):125203

[83] Xiong Q, Yang Z, Bruggeman P. Absolute OH density measurements in an atmospheric pressure dc glow discharge in air with water electrode by broadband UV absorption spectroscopy. Journal of Physics D: Applied Physics. 2015;**48**(42):424008

[84] Berden G, Engeln R. Cavity Ring-Down Spectroscopy: Techniques and Applications. New York: Wiley; 2009

[85] Zalicki P, Zare RN. Cavity ring-down spectroscopy for quantitative absorption measurements. The Journal of Chemical Physics. 1995;**102**(7): 2708-2717

[86] Stancu GD, Janda M, Kaddouri F, Lacoste DA, Laux CO. Time-resolved CRDS measurements of the $N_2(A^3\Sigma_u^+)$ density produced by nanosecond discharges in atmospheric pressure nitrogen and air. Journal of Physical Chemistry C. 2010;**114**(1):201-208

[87] Stancu GD, Kaddouri F, Lacoste DA, Laux CO. Atmospheric pressure plasma diagnostics by OES, CRDS and TALIF. Journal of Physics D: Applied Physics. 2010;**43**(12):124002

[88] Liu Z-W, Xu Y, Yang X-F, Zhu A-M, Zhao G-L, Wang W-G. Determination of the HO_2 radical in dielectric barrier discharge plasmas using near-infrared cavity ring-down spectroscopy. Journal of Physics D: Applied Physics. 2008;**41**(4):045203

[89] Gianella M, Reuter S, Press SA, Schmidt-Bleker A, Helden JH, Ritchie GAD. HO_2 reaction kinetics in an atmospheric pressure plasma jet determined by cavity ring-down

spectroscopy. Plasma Sources Science and Technology. 2018;**27**(9):095013

[90] Wang C, Wu W. Simultaneous measurements of OH(A) and OH(X) radicals in microwave plasma jet-assisted combustion of methane/air mixtures around the lean-burn limit using optical emission spectroscopy and cavity ringdown spectroscopy. Journal of Physics D: Applied Physics. 2013; **46**(46):464008

Chapter 2

Electrical Diagnostics of Dielectric Barrier Discharges

Floran Peeters and Tom Butterworth

Abstract

Atmospheric pressure dielectric barrier discharges (DBD) has many industrial applications and remains a focus of academic research. This chapter provides a thorough overview of electrical diagnostics for DBD, with a specific focus on charge-voltage measurement techniques. These methods are often underutilised in the existing scientific literature, despite the fact that they can provide useful insights into plasma behaviour. Both optimization of the electrical measurement setup and the interpretation of results are treated in-depth. The diagnostic techniques are discussed for a range of applications, from classic planar DBDs, to catalyst packed beds, plasma actuators, as well as techniques for measuring single microdischarges.

Keywords: electrical diagnostics, dielectric barrier discharges, packed bed, Q-V diagram, Lissajous figure

1. Introduction

Atmospheric pressure dielectric barrier discharges (DBD) has many industrial applications, due to its inherent simplicity and moderate operating conditions, and remains a focus of academic research. A schematic of a DBD is shown in **Figure 1**, along with a depiction of current and voltage characteristics over time. Upon breakdown of the gas in the gap, the dielectric prevents transition to a high current arc-like discharge, which would otherwise occur between two uncovered metal electrodes. Though spatially uniform discharging in DBDs can be achieved under certain conditions [1–7], under most circumstances a filamentary discharge will develop [8]. Since filaments are characterised by high local electron densities of up to 10^{15} cm^{-3} and strong electric fields of up to 10^5 V/cm, it is these filaments that determine the plasma chemistry [9–11]. The filaments are self-limiting, because they charge the dielectric surface and locally negate the gap voltage until extinction occurs within 10^{-7} s. Not only are the filaments spread out over the surface, they also occur over a wide timeframe. If, for example, the DBD is driven by a sinusoidal external voltage at a frequency of 100 kHz, filaments will ignite over a period of ≈3 μs in each half-cycle, at many different stages of the external voltage [12].

For many applications of DBD it is highly desirable to know the power dissipated in the discharge, e.g. for gas conversion or plasma actuators. The so called 'capacitor method' is generally the most accurate means of measuring plasma power, as well as allowing for more in-depth electrical characterisation of the reactor properties. For instance, the average gas gap voltage at which filaments (or any DBD plasmas in general) ignite can be determined with great accuracy using electrical diagnostics,

IntechOpen

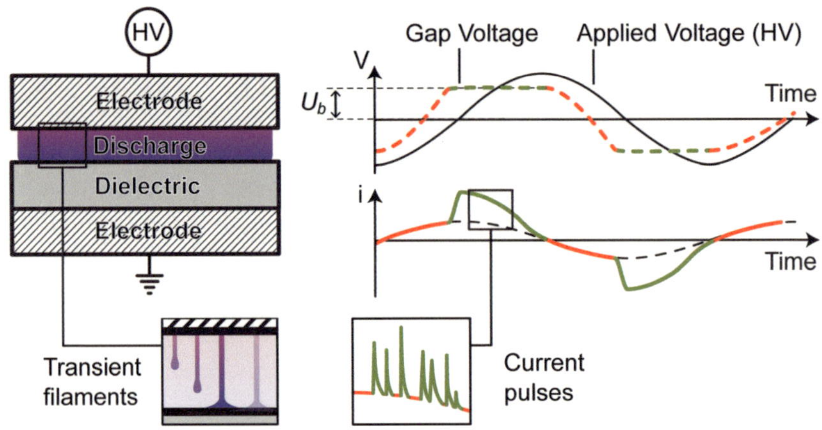

Figure 1.
Schematic of a DBD with one electrode covered by a dielectric. Current and voltage characteristics are given on the right-hand-side. While the plasma can be electrically modelled as having a continuous current and a constant gap voltage U_b *during discharging, in reality the plasma is composed of filaments with durations of* $< 10^{-7}$ *seconds separated in space and time.*

combined with a suitable equivalent circuit model. This ignition voltage tends to be constant and is here referred to as the 'burning voltage' U_b, as depicted in **Figure 1**. The burning voltage U_b provides a measure for the (reduced) electric field within the DBD plasma and is highly relevant for assessing the plasma chemistry occurring within the reactor [11, 13]. Along with large scale discharges, individual filaments have also been a subject of study, both experimentally [10, 14–16] and in modelling [9, 10, 17, 18]. In this regard, electrical diagnostics are ideally suited to assess the relationship between large-scale multi-filament discharges and computer simulations of individual filaments.

This chapter provides an overview of electrical diagnostics for DBD, with a focus on charge-voltage (Q-V) measurement techniques. Particular attention is paid to setting up Q-V measurements, identifying systematic errors, and performing in-depth analysis of the results to obtain reactor properties. Additionally, Q-V techniques for studying individual filaments and filament distributions are presented.

2. Charge-voltage diagrams in DBD

A charge-voltage (Q-V) diagram, or Lissajous figure, is obtained by making an X-Y plot of the voltage $V(t)$ across the DBD reactor's electrodes against the charge $Q(t)$ on a monitor capacitor. The voltage across the reactor electrodes is preferably measured using a calibrated high voltage probe (typically with a 1000:1 attenuation), connected as close as possible to the electrodes[1]. The monitor capacitor, having capacitance C_m, is situated between the reactor and the ground, and the voltage across it is measured using a standard (10:1 attenuation) oscilloscope probe. The instantaneous voltage $V_m(t)$ across the capacitor can be converted into the instantaneous charge on the capacitor $Q(t)$ by applying the relationship in Eq. (1). A schematic showing a typical experimental set-up to generate a Q-V diagram is shown in **Figure 2**.

[1]Alternatively, some power supplies feature an output voltage monitor. This method can be used, but should be treated with caution as a small phase difference in voltage and charge signals can lead to significant errors in power measurement.

$$Q(t) = C_m V_m(t) \tag{1}$$

2.1 Ideal Q-V diagrams

An ideal Q-V diagram can be defined as having a parallelogram shape, as shown in **Figure 3**. In a real DBD, there may be some deviations from this ideal shape, for which several cases are discussed in Section 3. The general principles behind analysis, however are the same for all Q-V diagrams. Parallelogram-shaped Q-V diagrams are the simplest case, where the discharging phases of the plasma reactor are clearly distinguishable by various line segments (AB, BC, CD and DA in **Figure 3**). Understanding the shape of the Q-V diagram, and deriving discharge properties from it, requires an equivalent electrical circuit, for which the simplest case is depicted in **Figure 4**. If we consider a parallel plate DBD during a full cycle of the applied voltage, there are two 'plasma off' phases (AB and CD in **Figure 3**) and two 'plasma

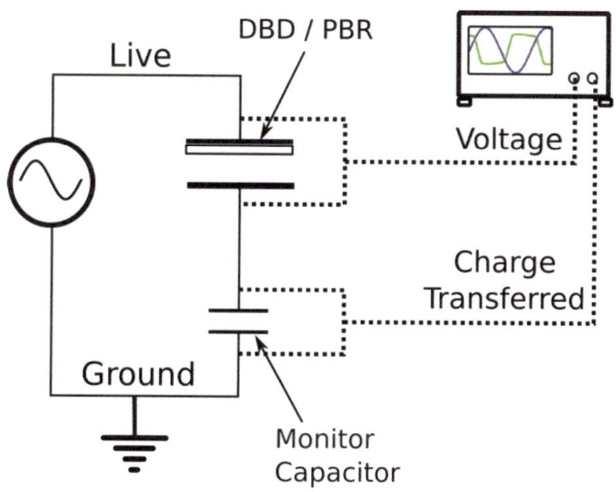

Figure 2.
Basic experimental schematic of a system used to generate a Q-V diagram using a dielectric barrier discharge (DBD) or packed bed reactor (PBR).

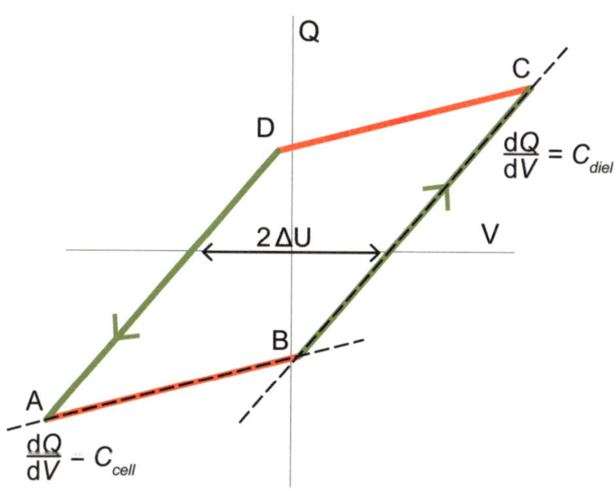

Figure 3.
Ideal *Q-V diagram with the characteristic parallelogram shape. 'Plasma off' segments are coloured in red, while 'plasma on' segments are green. Arrows in the green segments indicate the direction of time. Symbols are discussed in the main text.*

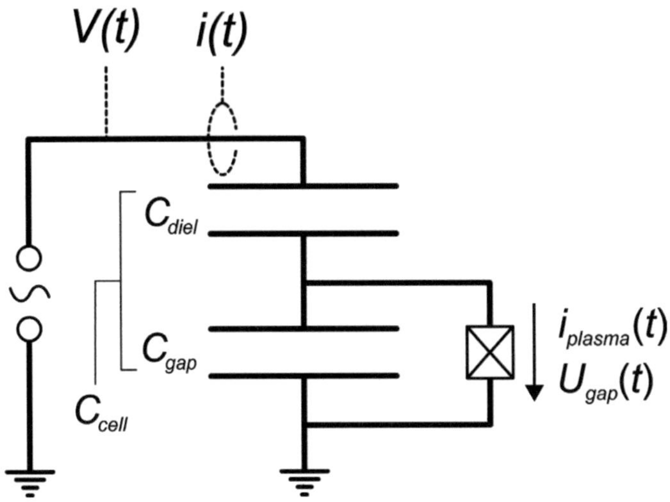

Figure 4.
The simplest equivalent electrical circuit corresponding to an ideal DBD. Note that the monitor capacitor with capacitance C_m *is not depicted in this circuit, as it is assumed to have a negligible effect on the electrical behaviour of the DBD.*

on' phases (BC and DA in **Figure 3**). During the 'plasma off', or capacitive phase, the reactor behaves as two capacitors in series. These capacitances are due to the dielectric layer, C_{diel}, and the discharge gas gap, C_{gap}, both being able to store charge, either from an applied voltage, or from charges accumulated by a plasma discharge. The overall 'DBD cell' capacitance, C_{cell}, of these two elements is found by applying Kirchoff's laws, yielding:

$$\frac{1}{C_{cell}} = \frac{1}{C_{diel}} + \frac{1}{C_{gap}} \tag{2}$$

Note that if both metal electrodes are covered with dielectrics, or if multiple gas gaps are found in series in the DBD reactor, the equivalent circuit of the reactor can be reduced to that depicted in **Figure 4**, where equivalent capacitances for both C_{diel} and C_{gap} can be found by applying a similar inverse addition as in Eq. (2).

The capacitances C_{cell} and C_{diel} can be obtained from the shape of an ideal Q-V diagram.[2] Remembering that $C = dQ/dV$, we can see that the gradients of the lines AB, BC, CD and DA in **Figure 3** correspond to constant capacitances. Lines AB and CD of the Q-V diagram correspond to the capacitive phase (i.e. 'plasma off'), where the gradient is equal to C_{cell}. This can be inferred from the equivalent circuit of **Figure 4**, since in the absence of plasma in the gap, current can only flow via the capacitive elements C_{diel} and C_{gap}.

Conversely, the gradient of the lines BC and DA corresponding to the discharging phase (i.e. 'plasma on') are usually identified as C_{diel} [19]. This is often (qualitatively) explained as the microdischarges 'bridging the gap', thus negating the C_{gap} term such that $C_{cell} = C_{diel}$ in Eq. (2). However, this view would imply that the voltage drop across the gas gap, U_{gap}, instantly drops to zero as soon as even a

[2]It is also possible to calculate approximate values for C_{diel} and C_{gap} for simple geometries. For a planar, parallel plate capacitor this is given by $C = \varepsilon_r \varepsilon_0 A/d$. Where ε_0 is the permittivity of free space, ε_r is the relative permittivity of the dielectric, A is area and d is dielectric thickness. In principal, this should give a reliable value of reactor capacitances, however, this method ignores edge effects which may be significant in some reactor geometries.

single microdischarge is ignited somewhere in the gap. The reasons for this are not immediately obvious based on inspection of **Figure 4**, and are only revealed after examination of the electrical behaviour of the discharge in the equivalent circuit.

Note that if no discharging would occur at all during a full cycle of the applied voltage $V(t)$ (if for instance the voltage amplitude is too low to achieve breakdown in the gas), the lines AB and CD would overlap and cross the origin in the Q-V plot. The fact that AB and CD develop opposing offsets on the Q-axis upon discharging is due to deposition of charges on the dielectric surface(s) by the plasma. In terms of the equivalent circuit, the time-varying (and surface-averaged) voltage across the gap capacitance $U_{gap}(t)$, as well as the voltage across the dielectric capacitance $U_{diel}(t) = V(t) - U_{gap}(t)$, become significantly modulated and phase-shifted with respect to the applied voltage $V(t)$, as is schematically depicted in the top-right-hand-side of **Figure 1**. Note the convention used here, where externally measurable voltages are referred to by V, and internal voltage drops are referred to by U (which are indirectly derived by modelling the electrical behaviour of the reactor).

2.2 Detailed analysis of the equivalent circuit

During the 'plasma on' phase, the gas gap in the reactor contains a variable, conducting medium, which is often represented by a time-dependent resistor $R(t)$, placed parallel to C_{gap} [20]. Since this approach requires $R(t)$ to be infinite until just before the 'plasma on' phase and neglects the possibility of an inductive element to the discharge, it is preferable to represent the discharge as a 'black box' element, through which a plasma current $i_{plasma}(t)$ flows due to the voltage $U_{gap}(t)$ across the gap [21]. In either case, the plasma current conductively transfers charge across the discharge gap, from a metal or dielectric coated electrode surface to an opposing dielectric electrode surface. As detailed in [21], the plasma current $i_{plasma}(t)$ and gap voltage $U_{gap}(t)$ for the equivalent circuit of **Figure 4** are given by:

$$i_{plasma}(t) = \frac{1}{1 - C_{cell}/C_{diel}} \left[\frac{dQ(t)}{dt} - C_{cell} \frac{dV(t)}{dt} \right] \quad \text{and} \tag{3}$$

$$U_{gap}(t) = V(t) - \frac{1}{C_{diel}} Q(t). \tag{4}$$

Eqs. (3) and (4) show how both current and voltage for the 'black box' element can be obtained from the measurable quantities C_{cell}, C_{diel}, $Q(t)$ and $V(t)$, without any a-priori assumptions regarding the electrical behaviour of the plasma. The term between brackets in Eq. (3) consists of the externally measured current, $dQ(t)/dt = i(t)$, from which a capacitive current, $C_{cell} \cdot dV(t)/dt$ is subtracted. This last term is often erroneously referred to as the *displacement current* of the DBD. The true displacement current, defined as $i_{displacement}(t) = i(t) - i_{plasma}(t)$, can be quite different due to the pre-factor $1/(1 - C_{cell}/C_{diel})$ in Eq. (3). Neglecting this pre-factor can lead to underestimation of $i_{plasma}(t)$ and overestimation of $i_{displacement}(t)$, especially in reactors where the ratio $C_{gap}/C_{diel} \gg 0.01$. Physically, this pre-factor is due to the electrical energy stored in C_{gap} prior to discharge ignition, which is subsequently drained through the plasma and leading to an enhanced plasma current during discharging.

The equivalent circuit model implicitly assumes a spatially uniform discharge, so that gap voltage $U_{gap}(t)$, described by Eq. (4), is always an average over the DBD surface area. This is worth keeping in mind when interpreting results on the level of individual microdischarges, but is certainly sufficient to describe reactor-averaged properties resulting from large numbers of filamentary microdischarges.

The behaviour of $U_{gap}(t)$ can reveal an important feature of the equivalent circuit in **Figure 4** after solving for $Q(t)$ and differentiating with respect to $V(t)$:

$$\frac{dQ(t)}{dV(t)} = C_{diel}\left(1 - \frac{dU_{gap}(t)}{dV(t)}\right). \tag{5}$$

Eq. (5) is always valid for the equivalent circuit in **Figure 4** (also during 'plasma off' phases), irrespective of the shape of the Q-V diagram it is assumed to describe. For a parallelogram-shaped Q-V diagram, however, the term between brackets must be a constant, or AB and CD, or BC and DA, would not form straight lines. Identification of the gradients of BC and DA with C_{diel}, as is usual in the literature, requires that $dU_{gap}(t)/dV(t) = 0$ during a discharge half-cycle (reducing the term between brackets in Eq. (5) to (1). Since the applied voltage $V(t)$ changes significantly over the time-span of a 'plasma on' phase (i.e. $dV(t) \neq 0$), this implies that $dU_{gap}(t) = 0$ while there is discharging occurring in the gap. In other words, a straight line in a Q-V diagram during a 'plasma on' phase indicates a constant U_{gap} from the moment the discharge begins, until its end. See also **Figure 1** for an illustration of this phenomenon, where this constant gap voltage is identified as the burning voltage U_b.

Eq. (5) is applicable to both 'plasma on' and 'plasma off' phases. The straight lines AB and CD during 'plasma off' imply that Eq. (5) must equal C_{cell}, meaning that the term between brackets is a constant with a value of C_{cell}/C_{diel} during these periods. During 'plasma on', straight lines emerge as long as the term $dU_{gap}(t)/dV(t)$ is a constant, but not necessarily equal to zero. In general, Eq. (5) can be used to derive the relationships between external voltage $V(t)$ and gap voltage $U_{gap}(t)$ for parallelogram-shaped Q-V diagrams:

$$\left.\frac{dU_{gap}}{dV}\right|_{off} = 1 - \frac{C_{cell}}{C_{diel}} = \frac{C_{diel}}{C_{gap} + C_{diel}} \quad \text{and}$$
$$0 \leq \left.\frac{dU_{gap}}{dV}\right|_{on} < \frac{C_{diel}}{C_{gap} + C_{diel}}, \tag{6}$$

where the time-dependence in dU_{gap}/dV has been removed to indicate that this rate is a constant during either 'on' or 'off' phases. The equality for 'plasma off' in Eq. (6) matches Kirchhoff's laws for the division of voltage between two capacitors in series. It further shows that U_{gap} responds more strongly to changes in external voltage V if $C_{diel} \gg C_{gap}$. As will be shown in Section 3.5, this is relevant to quickly regain small drops in U_{gap} occurring between individual filamentary discharges. Meeting the condition $C_{diel} \gg C_{gap}$ ensures a maximum number of filaments is ignited during a 'plasma on' period, since any reduction in U_{gap} due to discharging is more quickly compensated by the external voltage V. Regarding the inequality in Eq. (6) for the 'plasma on' phase, there is no a-priori reason to assume that $dU_{gap}/dV = 0$, since any constant value for dU_{gap}/dV can produce a constant dQ/dV gradient during 'plasma on' in a parallelogram-shaped Q-V diagram. As Eq. (5) already showed, C_{diel} simply represents the highest constant dQ/dV gradient that can be observed in an ideal DBD, corresponding to $dU_{gap}/dV = 0$, i.e. a constant gap voltage U_b during discharging. Values for $dQ/dV < C_{diel}$, and a non-zero dU_{gap}/dV, during 'plasma on' are equally possible. This situation occurs in the case of partial surface discharging, where only a portion of the available dielectric surface area is exposed to discharges during 'plasma on' phases, and will be discussed further in Section 3.1. The inequality in Eq. (6) becomes invalid for individual microdischarges, since these do not produce ideal parallelogram Q-V diagrams. In these cases dQ/dV gradients can easily exceed C_{diel}, which is related to the sudden drops in gap voltage produced by microdischarges; a situation which is discussed further in Section 3.5.

2.3 Power measurements

The most common usage of a Q-V diagram is to determine the *active* power dissipated in the discharge using the area covered by the hysteresis loop [22]. In some of the available literature, the equations derived for the power calculation are based upon the geometric features of the Q-V diagram by assuming a perfect parallelogram [19, 23]. In practice, as will be demonstrated in Section 3, this ideal behaviour rarely occurs. For completion, the relationship between area and average power will be derived here, valid for any shape of the Q-V diagram. Differentiating Eq. (1) with respect to time, the current through the monitor capacitor $i_m(t)$ becomes:

$$i_m(t) = C_m \frac{dV_m(t)}{dt} = i(t), \tag{7}$$

where the last equality is due to continuity of current through the series circuit of DBD + monitor capacitor in **Figure 2**. The instantaneous power $P(t)$ in the DBD is then given by:

$$P(t) = V(t) \cdot i(t) = V(t) \cdot C_m \frac{dV_m(t)}{dt}. \tag{8}$$

The time-averaged power $\overline{P}$ can be calculated by integrating Eq. (8) with respect to time and substituting the charge relationship of Eq. (1):

$$\overline{P} = \frac{1}{T}\int_0^T V(t) \cdot C_m \frac{dV_m(t)}{dt} dt = \frac{1}{T}\int_0^T V(t) \cdot C_m dV_m(t) = \frac{1}{T}\oint V(t) dQ(t), \tag{9}$$

which shows that the average power dissipated in a full discharge cycle is given by the area bounded by the Q-V diagram, multiplied by the frequency of the discharge cycle. When using numerical methods to calculate this integral, it is imperative to ensure integration of acquired data is performed over exact multiples of the discharge time T, or the *reactive* power in the capacitive parts of the circuit will not average out to zero.

2.4 Selection of an appropriate monitor capacitor

In order to apply the Q-V method, a suitable capacitor must be selected. The required monitor capacitor is dependent upon the capacitance of the reactor C_{cell}. The capacitance of the monitor capacitor should greatly exceed the capacitance of the reactor, i.e. $C_m \gg C_{cell}$. As a rule of thumb, a ratio of C_m:C_{cell} between 100:1 to 10,000:1 is generally appropriate. If the selected monitor capacitance is too small, the voltage across the capacitor will exceed the maximum voltage range of the oscilloscope (for a typical oscilloscope this is 20 V full scale), and the signal will over-range. The voltage on the monitor capacitor, *in the absence of discharging in the DBD*, will be:

$$V_m(t) = V(t)\frac{C_{cell}}{C_m}. \tag{10}$$

Taking into account the common 10:1 attenuation for an oscilloscope probe, Eq. (10) divided by a further factor 10 provides a lower limit on the signals to be detected by the oscilloscope. As a rule-of-thumb, the peak signal to be detected *during discharging* can exceed this lower limit by a factor of 5–10, though this depends strongly on the configuration of the DBD. Typically, the greater the ratio between dielectric capacitance C_{diel} and gap capacitance C_{gap}, the greater the amplitude of $Q(t)$, and thereby $V_m(t)$, will become during a full period.

Eq. (2) also shows that when C_m is too large compared to C_{cell}, the amplitude of the voltage across the monitor capacitor will become very small, and the resulting low signal-to-noise ratio will become problematic (a typical oscilloscope has a minimum range of 20 mV full scale, but also a non-zero noise floor in the 1 mV range). It is also worth noting that most oscilloscopes use 8-bit analog-to-digital converters (ADCs), which divides the full scale into 256 discrete steps. For accurate measurement of $V_m(t)$, and thereby $Q(t)$, optimising the value of C_m to make as much use of the oscilloscope's full scale is recommended. Depending on the maximum values of $V_m(t)$, it is possible to forego the use of 10:1 oscilloscope probe and directly attach the poles of the measurement capacitor to a coaxial line to the oscilloscope. This should only be attempted if $V_m(t) \lesssim 3$ V, since the oscilloscope channel will have to be set to 50 Ω to prevent reflections in the line, which simultaneously has a low maximum DC voltage limit (usually < 5 V). The advantage of this method is a high bandwidth and a higher signal-to-noise than oscilloscope probes can provide.

Provided $C_m \gg C_{cell}$, almost all the voltage applied to the reactor + monitor capacitor drops across the DBD, keeping the equivalent circuit in **Figure 4** valid, while not subjecting the monitor capacitor to high voltages. Therefore, a relatively low voltage capacitor in the range of 100–250 V may be selected (a value motivated more by a desire for robustness than with an expectation of applying 100 V to it during measurements). With regards to the type of capacitor, class 1 ceramic capacitors are preferred, as they have relatively low parasitic inductances and parasitic capacitances, providing a greater accuracy and stability with resonant circuits (of which DBD reactors, as well as the measurement circuit, are examples). If possible, it is worth measuring the capacitance of the monitor capacitor and the assembled monitoring circuit with an LCR meter, as measured values can fluctuate from the values given by the manufacturer. The monitoring circuit can also introduce additional capacitance, leading to additional inaccuracies in determining the absolute value of $Q(t)$.

The voltage across a shunt resistor, or an inductive Rogowski coil, can also be used in place of a monitor capacitor in **Figure 2**. This method relies on direct measurement of current $i(t)$ from DBD to ground, instead of integrating the current on a capacitor. Time-integration of $i(t)$ can then be employed in post-processing to obtain $Q(t)$. Since filamentary discharges coincide with fast-rising currents in the sub-nanosecond range, this approach ideally requires a high bandwidth measurement system and a high sampling rate to prevent measurement errors. The advantage of integrating on a capacitor is essentially that no charge moving through the reactor can be 'missed', even with low bandwidth (and sampling rate) of the probe + oscilloscope, allowing for e.g. more accurate power measurements.

As a final note, certain reactor geometries have a poorly defined ground electrode, making it difficult to attach a monitor capacitor, shunt resistor, or coil. The 'ground' electrode can be electrically floating, for instance, or multiple potential pathways can exist for current to reach ground, such as with plasma jets impinging on a surface. In these cases, it may be beneficial to use a Rogowski coil after all, and measure current directly on the high voltage line between voltage source and reactor. This poses additional engineering and safety challenges, however, since unwanted discharging, or accidental application of high voltages to measurement devices must be prevented. While caution is advised, these problems can be overcome. Care must be taken to prevent sharp metallic edges from forming corona discharges, and air-gaps between high voltage line and coil from forming arc discharges. This can be achieved by covering and filling in all sharp edges and air gaps with a higher breakdown strength material, such as e.g. paraffin. as was done in [24].

3. Non-ideal DBD scenarios

There are a number of scenarios in which Q-V diagrams do not display a parallelogram shape. This can be due to the real, physical characteristics of the reactor, or experimental error. In this section, various shapes of Q-V diagrams will be discussed, along with the possible implications regarding the characterisation of the reactor.

The following deviations from ideal behaviour will be addressed:

1. Q-V diagrams with gradients less than C_{diel}

2. Almond shaped Q-V diagrams, with rounded corners at the transition from non-discharging to discharging regions

3. Elliptical Q-V diagrams, with no straight lines at all

4. Noisy Q-V diagrams, particularly during 'plasma on' phases

5. Stepped Q-V diagrams

This list is not exhaustive, but should provide a troubleshooting guide for any problems that arise, both experimentally and in interpretation of data. The fundamentals for the analysis of these plots are all based on the equivalent circuit of **Figure 4**, combined with the theory provided in the previous sections.

3.1 Q-V diagrams with gradients less than C_{diel}

A common irregularity observed in Q-V diagrams for DBDs and PBRs is a gradient during discharging changes depending on experimental conditions, despite the geometry of the reactor remaining the same. This is caused by partial discharging of the available (dielectric) area in the gas gap. Partial discharging is commonly observed when the voltage applied to the discharge gap is not far in excess of the reactor breakdown voltage [21, 25, 26]. It is also a common feature when dielectric packing materials are introduced into the gas gap, as is the case in PBRs, which lead to a complicated (electrical) geometry with a range of gap widths and breakdown voltages [13, 27–30]. The latter case will be treated in the next section, while this section focuses on parallelogram-like Q-V diagrams resulting from plane-parallel reactor geometries.

Example Q-V diagrams where changing the applied voltage amplitude leads to different gradients in the discharging phase are shown in **Figure 5**. Aside from applied voltage amplitude, changing the fractional composition of the gas in the

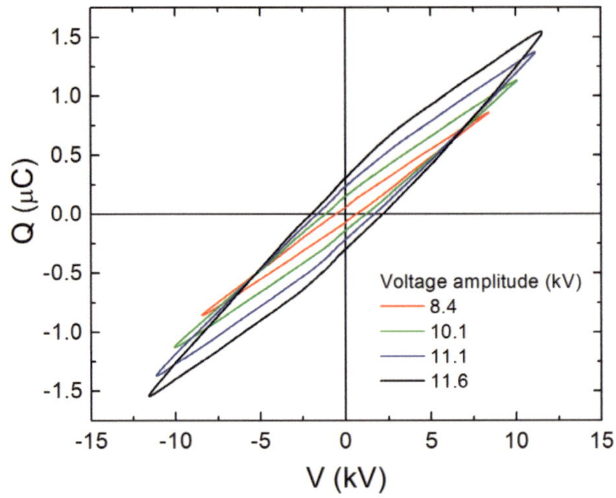

Figure 5.
Example Q-V diagrams where increasing the applied voltage amplitudes leads to steeper gradients during 'plasma on' phases. The 55 kHz plane-parallel DBD in this example uses N_2 gas and has a high C_{gap}/C_{diel} ratio, with $C_{gap} \approx 240$ pF and $C_{diel} \approx 160$ pF.

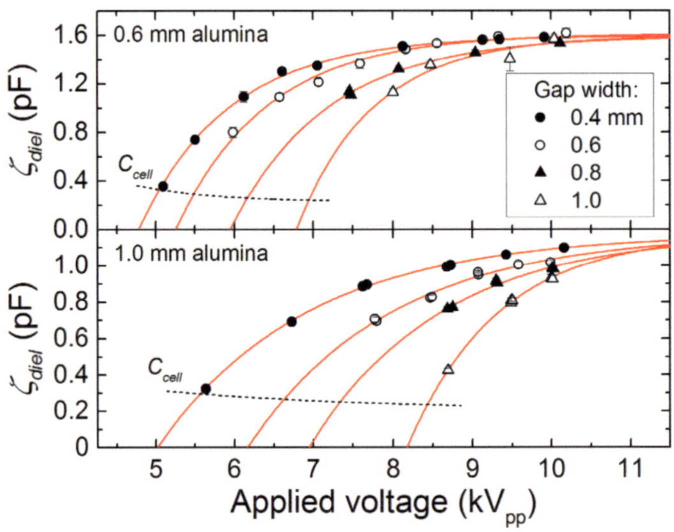

Figure 6.
Effective capacitance during 'plasma on' phases ζ_{diel} for 8 different 100 kHz plane-parallel DBD configurations in air as a function of applied voltage amplitude. The red lines are fits of single exponential functions saturating to the same value for C_{diel} each of the two dielectric thicknesses. The dotted lines indicate the cell capacitance C_{cell} for a given configuration.

discharge, or changing the frequency of the applied voltage, can have similar effects [31]. In these cases, the gradient of the 'plasma on' phase is termed the effective dielectric capacitance ζ_{diel}, to distinguish it from the dielectric capacitance C_{diel}, which should only depend on the geometry of the reactor. The fact that the gradient during 'plasma on' phases can be less than, but never exceed, C_{diel} was already discussed in Section 2.2. Examples of changes to ζ_{diel} as a function of gap width, dielectric thickness and applied voltage are shown in **Figure 6**.

This data demonstrates that the higher the C_{gap}/C_{diel} ratio (corresponding to narrower gas gaps, and thicker dielectrics and/or lower permittivity of the dielectric material), the broader the range of applied voltage amplitudes over which partial surface discharging will occur. It also demonstrates that the value of C_{diel} to be used in the equivalent circuit of **Figure 4**, and any subsequent analysis of *Q-V* diagrams, can simply be defined as the value towards which ζ_{diel} saturates at sufficiently high applied voltage amplitudes. A method to characterise the properties of the reactor in the intermediate range where $\zeta_{diel} < C_{diel}$, including the fractional discharged area, are described in detail in [21].

3.2 Almond shaped Q-V diagrams

Almond shaped Q-V diagrams, an example of which is depicted **Figure 7**, can be observed in PBRs, particularly with high dielectric constant materials [31], and with surface discharge DBDs such as plasma actuators [32]. In both cases, the almond shape is caused by the gradual expansion of the plasma across the available dielectric area during the discharge phase. In a plane-parallel DBD configuration, the gas gap is uniformly spaced and the plasma ignition voltage across the gap, the burning voltage U_b, is approximately constant throughout the discharging phase. As pointed out in Section 2.2, this constant voltage across the gap is a requirement for observing straight lines in Q-V diagrams during the discharging phase. In discharges with non-uniform gap widths, plasma will ignite across a wide range of gap voltages, since the breakdown criterion is not the same everywhere.

In PBRs, the electric field strength prior to discharging is highest at certain localised points in the gap, either in the small gas gaps between, or near sharp features of, packing particles [33, 34]. In a plasma actuator, see **Figure 8**, the electric field is initially strongest at the minimum distance between the two electrodes. In both cases, as the applied voltage increases during the AC cycle, discharging will commence at the point where the electric field strength is highest (and the breakdown criterion is met first) and then gradually expand across the surface of the electrode, as the breakdown criterion is met for wider and wider gaps. Whilst this plasma expansion occurs, the fractional surface area of the gap capacitance C_{gap} and dielectric capacitance C_{diel} actually participating in the discharge increases, leading to an effective capacitance during discharging ζ_{diel} which grows during a half-cycle, see **Figure 8**. When the dielectric area charged by the plasma reaches a maximum, the gradient of the line will be at its steepest. Similar to the 'ideal' DBD discharges, once the applied voltage $V(t)$ reaches its maximum no further discharging can occur, because electric field strengths will only start

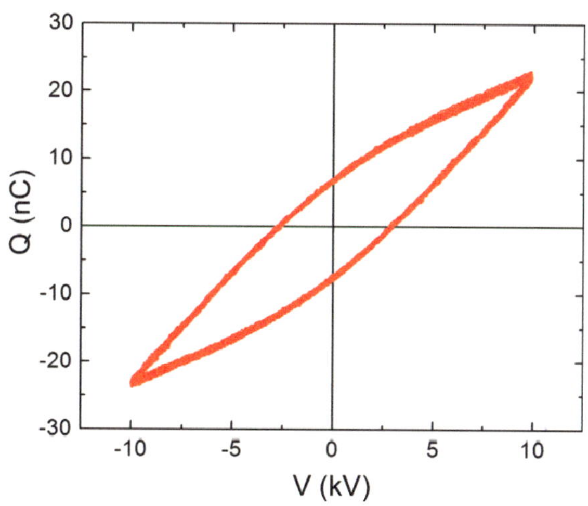

Figure 7.
'Almond'-shaped Q-V diagram generated using a $BaTiO_3$ packed bed DBD discharging in 90% Ar–10% CO_2. The reactor is the same as that used in [31].

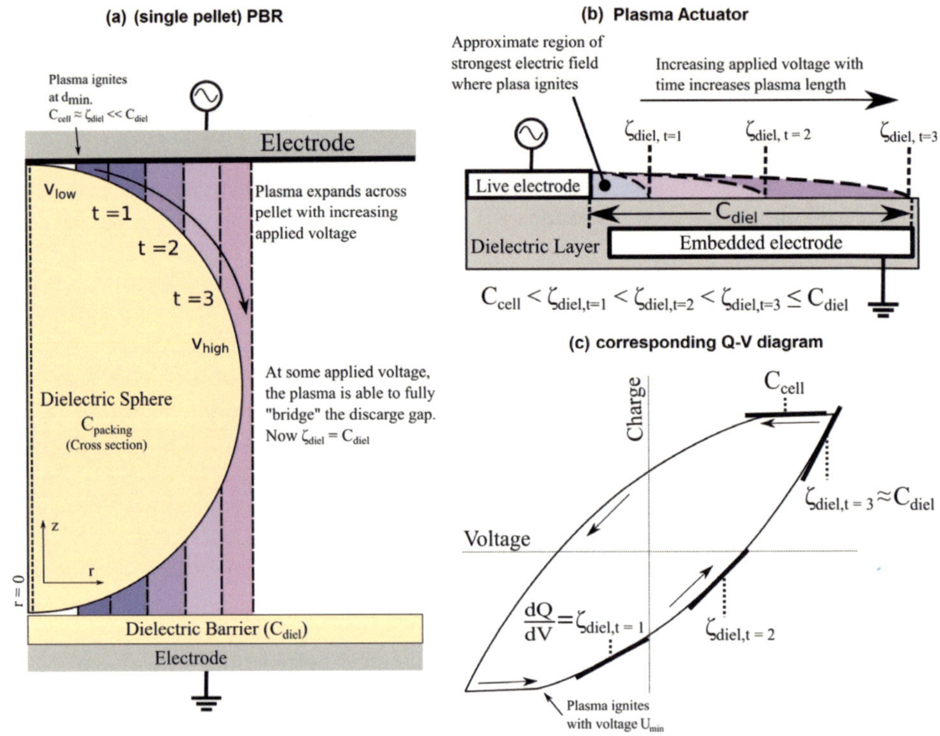

Figure 8.
Plasma expansion across (a) a single pellet PBR and (b) a DBD plasma actuator, and (c) its influence on the shape of the Q-V diagram.

decreasing everywhere in the gas gap and no further breakdown criteria will be met. This leads to the sharp apex at the maximum and minimum voltage and charge amplitudes on the Q-V diagram.

Regarding the analysis of almond shaped Q-V diagrams, the theory is the same as for Q-V diagrams with gradients less than C_{diel} discussed in the previous section. The difference being that the effective capacitance ζ_{diel} is now a function of time. In a similar fashion, C_{diel} can be determined from the maximum observed slope (i.e. ζ_{diel3} in **Figure 8**), preferably from Q-V diagram obtained at the highest attainable applied voltage amplitude.

Seen from the perspective of the equivalent circuit of **Figure 4**, what needs to be kept in mind is that the entire area of the reactor that can potentially discharge is treated as a single point. This means that $i_{plasma}(t)$ and $U_{gap}(t)$ obtained via the equivalent circuit represent instantaneous values, averaged over the entire reactor area. Especially in packed beds, where the true electrical geometry is far more complicated than the simplified DBD equivalent circuit, no detailed, local information on the discharge properties can be obtained. Nonetheless, reactor-averaged values for e.g. $U_{gap}(t)$ and effective surface coverage of the discharge can be determined and effectively compared between different reactor geometries and experimental conditions. See for example [31, 32], where it is also explained how the equivalent circuit in **Figure 4** can be amended to include the capacitance of the packing materials.

3.3 Elliptical Q-V diagrams

Elliptical Q-V diagrams have been observed in a number of scenarios [35, 36]. For elliptical Q-V diagrams to be physical, there would have to be a continuous plasma current $i_{plasma}(t)$ during all phases of the discharge cycle. The elliptical shape is then indicative of a significant residual electron density in the discharge gap surviving through the 'plasma off' period between sequential (filamentary) discharge cycles. Below the possibility of this occurring will be investigated.

Referring once more to the equivalent circuit in **Figure 4**, the typical capacitive behaviour during a 'plasma off' period would cease if the 'black box' element representing the plasma maintains an effective residual impedance $R_{residual}$ lower than the impedance of the gap capacitance C_{gap}:

$$R_{residual} \ll \left|\frac{1}{i\omega C_{gap}}\right|, \tag{11}$$

where $\omega = 2\pi f$, with f the driving frequency of the applied voltage $V(t)$. Using this criterion, the maximum impedance of residual volume plasma leading to an elliptical Q-V diagram can be determined. With typical electron mobilities at atmospheric pressure $\mu_e \approx 10^{-1}$ m^2/Vs [37, 38] versus positive or negative ion mobilities $\mu_{ion} < 10^{-3}$ m^2/Vs [39–41], we will assume for now that electrons determine the conductivity. Further limiting the analysis to plane-parallel DBD configurations with gap width d_{gap} and electrode area $A_{discharge}$, we can write:

$$C_{gap} = \frac{\varepsilon_0 A_{\text{discharge}}}{d_{gap}} \quad \text{and} \tag{12}$$

$$R_{residual} = \rho_{residual}\frac{d_{gap}}{A_{\text{discharge}}} = \frac{1}{e\mu_e n_{e,\,res}}\frac{d_{gap}}{A_{\text{discharge}}}, \tag{13}$$

with $\rho_{residual}$ the resistivity of the residual plasma, e the electron charge, $n_{e,res}$ the residual electron density and ε_0 the permittivity of free space. By substituting Eqs. (12) and (13) in (11) and solving for $n_{e,res}$ we find the minimum gas-phase electron density required to have a noticeable plasma current throughout a full period of the applied voltage:

$$n_{e,\,res} \gg \frac{\varepsilon_0 2\pi f}{e\mu_e} \approx f \cdot 10^{10}\text{m}^{-3}. \tag{14}$$

This result indicates that the minimum required residual electron density does not depend on DBD geometry, only on the driving frequency of the applied voltage f and the mobility of charge carriers in the gas-phase.

The next question is what residual electron density can be sustained between discharge half-cycles. To estimate this, the approximate recombination time of residual electron-ion pairs $\tau_{e,res}$ at atmospheric pressure and room temperature can be compared to the driving frequency:

$$f \approx \frac{1}{\tau_{e,\,res}} = k_{rec} n_{e,\,res} \approx n_{e,\,res} \cdot 10^{-12}\text{m}^3\text{s}^{-1}, \tag{15}$$

where a typical value for the electron-ion recombination rate k_{rec} is used [42]. According to Eq. (15), at a typical frequency f = 100 kHz, a residual electron density $n_{e,res} \approx 10^{17}$ m^{-3} (ionisation degree < 10^{-8}) can be sustained. More detailed models, including losses of charged species to (dielectric) surfaces, predict $n_{e,res} \gtrsim 5\times10^{17}$ m^{-3} under these conditions [43]. When compared with Eq. (14) this is sufficient to expect a Q-V diagram to have a noticeably elliptical shape. In fact, this is true for all frequencies, as well as if we assume conductivity to be determined by ion mobility instead of electron mobility. Why then are elliptical Q-V diagrams not consistently observed in DBD? The answer is that while $i_{plasma}(t)$ always remains greater than zero during 'plasma off' phases, it is negligibly small compared to the large plasma currents occurring during 'plasma on' phases. The tendency for Q-V diagrams to become more elliptical, especially with increasing power can, however, be observed as small increases to the 'plasma off' gradient [24, 44, 45]. This apparent increase in the C_{cell} capacitance can be attributed to the formation of plasma sheaths between the weak residual plasma in the centre of the gas gap and the surrounding surfaces, which can be treated as a small (and likely time-dependent) additional capacitance parallel to C_{gap} in the equivalent circuit. This is why C_{cell} is best determined from Q-V diagrams measured at low applied voltage amplitudes, ensuring that no discharging occurs.

Truly elliptical Q-V diagrams can be expected if ionisation (and increased electron densities) can be sustained over full periods of the AC voltage, but this is only expected for capacitive discharges at gas pressures well below 1 bar [46]. If elliptical Q-V diagrams are observed at atmospheric pressure, it could (a) be the result of a relatively unstable discharge combined with averaging Q-V data over a large number of applied voltage cycles, or (b) an error in the measurement setup.

A good way to check the measurement setup is to run it at low voltages, when no plasma is present in the reactor. If an elliptical Q-V diagram emerges, it can be indicative of a parasitic capacitance or inductance. A common source of parasitic components to the measurement circuit is poor connections between monitor capacitor, probe and/or oscilloscope. This can have an impact on the measurement of plasma power and the characterisation of the reactor when it is in operation, since the Q-V diagram may already have a non-zero area in the absence of plasma. This is especially important with small reactors, or applications where a high sensitivity power measurement is required [28]. The phase angle of an inductive circuit is different to a capacitive circuit, so that the phase shift between the voltage across the monitor capacitor $V_m(t)$ relative to the applied voltage $V(t)$ can be used to identify potential problems, or simply to assess the accuracy of any subsequent power measurements. Note that these phase shifts (and elliptical Q-V diagrams) can also emerge if the measurement delay between the high voltage probe for V(t) and the voltage probe for Q(t) is significant. Since 1 meter of coaxial cable will provide only a $\approx$ 3–4 ns delay, this is not a factor worth considering for large scale DBDs, but it does become relevant for electrical measurements requiring a high time-resolution, such as for single filament discharges (**Figure 9**).

3.4 Noisy Q-V diagrams

It is a common, and almost unavoidable, occurrence for Q-V diagrams to show noise during the discharging phase of the DBD cycle. The noise is caused by interference from the plasma microdischarges themselves, hence the reason that they only occur during the discharging phases. An example of noise experienced in Q-V diagrams is shown in **Figure 10**, as well as Refs. [47,48]. The unfiltered plot in **Figure 10** shows background noise during the capacitive phase, but larger spikes during the discharging phase of the DBD cycle.

The background noise is usually broadband in nature and, provided it is not due to measuring near the noise floor of the oscilloscope (see Section 2.4), can be attributed to RF pickup. To mitigate this type of noise it is essential to ensure the shielding on both ends of a coaxial line is connected to the same ground. Moreover, when using low voltage oscilloscope probes, it is good practice not to use the alligator clip ground wire supplied with the probe, but to solder one end of a short wire directly between the ground side of the monitor capacitor and wrapping the other end tightly around the grounded barrel of the probe. The grounded barrel of

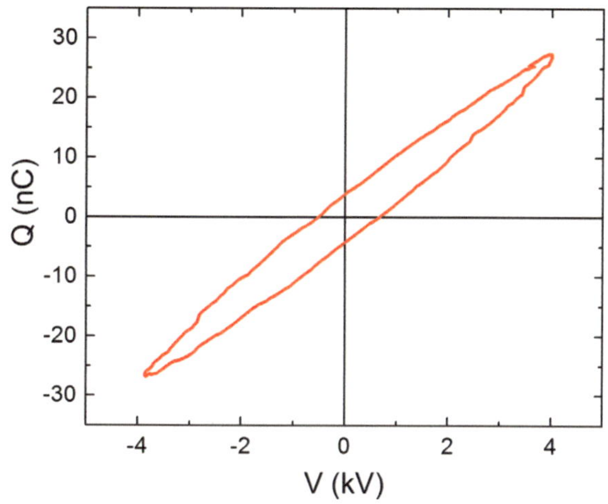

Figure 9.
Elliptical Q-V diagram where stray inductance and capacitance are the dominating influence of the plot shape. Data is obtained using an unpublished coaxial PBR. The steps in the discharging line show the presence of streamer discharges, however, the curved capacitive line indicates that stray capacitance is influencing the Q-V diagram shape.

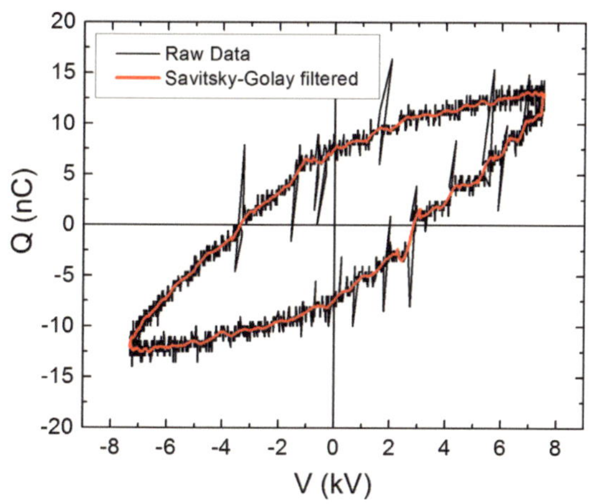

Figure 10.
Q-V diagram of unfiltered, noisy data, alongside the same data that has had the noise removed using a 2nd order, 20 point Savitzky-Golay filter. The original data is obtained using a small electrode area DBD with a dielectric packing material.

the probe may be covered by a plastic sheath, but this can usually be removed with ease. The concept behind this approach is to remove any undesirable 'loop antennas' from the circuit, which would otherwise pick up any radiated electromagnetic noise from both the DBD reactor and other nearby sources.

The large spikes are a direct consequence of the rapid increase in current (and correspondingly rapid shifts in charge) in the reactor + measurement system due to the ignition of individual filaments. With current rise times of < 1 ns, the individual discharges introduce high frequency harmonics into the measurement circuit. This becomes visible on the oscilloscope as 'ringing' signals, such as depicted in **Figure 11**. Generally, this 'ringing' is caused by impedance discontinuities in the measurement circuit. In any coaxial transmission line, transmitted voltage waves will reflect to some degree at points where the impedance is discontinuous. These impedance discontinuities tend to occur where the coaxial line is attached to the monitor capacitor, or where the line is attached to the scope, and will introduce at

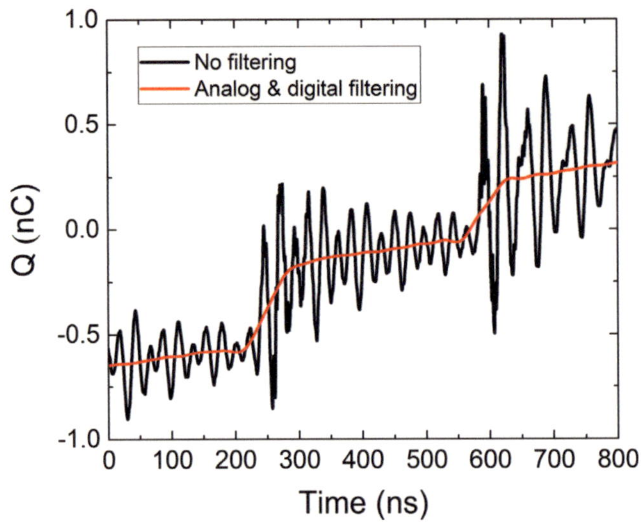

Figure 11.
Close-up of a ringing signal in Q(t) related to single filamentary discharges and a low pass filtered Q(t) signal.

least some parasitic inductance. Since the value of this parasitic inductance L depends on the physical length of the discontinuity, with higher frequencies producing a higher L for a given length, the impedance created by any given discontinuity is proportional to f^2 and becomes significant for high frequency signals. The steep rising edge signals from a filamentary current pulse will then appear as in **Figure 11**, where a single pulse produces damped oscillations due to the R-L-C circuit formed by the measurement system. To reduce this parasitic inductance on the reactor side (which is likely the dominant source), best practice is to attach the probe head *and* ground line (or the core and shielding of the coaxial line, if no probe is used) as close as possible to the electrodes of the monitor capacitor. The number of coaxial connectors is also best kept to a minimum, i.e. using a single cable is recommended.

Note that parasitic inductance, and subsequent ringing, is always introduced wherever a coaxial transmission line is interrupted, irrespective of whether the connections are mechanical or soldered. A standard method of removing the ringing is by including a low pass filter in the circuit. It is desirable to do this in the measurement circuit itself and not in post-processing, since ringing can introduce significant overshoots in the (apparent) measured voltage $V_m(t)$, making it impossible to make full use of the vertical resolution of the oscilloscope. A capacitor C in series with a resistor R, placed between the electrodes of the monitor capacitor, with the probe electrodes attached to the filter capacitor will provide a cut-off frequency $f_{LP} = (RC)^{-1}$. Common values for f_{LP} are 50 MHz, $R = 100\ \Omega$ and $C = 200$ pF. This method will, however, reduce the bandwidth of the measurement and will reduce signal magnitudes of $V_m(t)$ to some degree, introducing errors into e.g. power measurements. To determine the highest possible cut-off frequency for low pass filtering, performing a Fast Fourier Transform (FFT) of the unfiltered measured signal can be considered [49]. If it is of particular interest to measure the rise time of a single filamentary discharge, the loss of bandwidth introduced by filtering may be undesirable. It is worth keeping in mind, however, that this rise time may be too short to readily measure with standard equipment in any case [50]. Additional noise removal can be done in post-processing using the Savitzky-Golay algorithm [51]. This is a commonly used averaging technique that provides satisfactory noise reduction whilst simultaneously minimising the loss of signal magnitude.

3.5 Stepped Q-V diagrams

The stepped Q-V diagram typically occurs when a small electrode area is used, making the number of individual streamers that can be formed per AC half-cycle low enough for individual filaments to be observed. In this case each 'step' in the Q-V diagram corresponds to an individual filamentary streamer transferring charge between the electrodes, see **Figure 12** for an example. Stepped Q-V diagrams cause difficulty in trying to determine certain reactor operating characteristics from the geometry of the Q-V diagram, particularly burning voltage U_b and the maximum capacitance C_{diel}. The nearly horizontal segments between each step have a gradient close to C_{cell}, since very little plasma current $i_{plasma}(t)$ is flowing between filament ignitions. The steep, almost vertical part of each step in **Figure 12**, is the point where $i_{plasma}(t)$ is highest, with a gradient that is clearly higher than the average slope along the line BC. This high gradient violates the statement in Section 2.2 that the slope in a Q-V diagram can never exceed C_{diel}. However, this statement is valid only for parallelogram-shaped Q-V diagrams, and assumes the discrete nature of individual discharges can be neglected. Instead, Eq. (5) can be used to interpret the nearly vertical slopes: during these instances dU_{gap}/dV would have to be less than zero for dQ/dV to exceed C_{diel}. In other words, during a filamentary discharge, the *area-averaged* gap voltage $U_{gap}(t)$ in the reactor must drop faster than the applied voltage $V(t)$ increases. This is exactly what occurs during the nearly vertical gradients seen in **Figure 12**. In practice, however, $V(t)$ also decreases during a single discharge (i.e. $dV < 0$), because real voltage sources are not ideal. The effect of $dV < 0$ during discharging can also be seen in a number of vertical steps in **Figure 12** as negative gradients on the Q-V diagram (sloped from bottom right to top left). Since this implies that $dQ/dV < 0$, Eq. (5) suggests that $dU_{gap}/dV > 1$ at these points. Since dV is negative, dU_{gap} must also be negative and have a greater magnitude than dV. Both of these features can be observed in **Figure 13**, where $i_{plasma}(t)$ and $U_{gap}(t)$ have been calculated from the data in **Figure 12** using the equivalent circuit approach.

To utilise the equivalent circuit of **Figure 4** to determine (reactor-averaged) discharge properties from Q-V diagrams, as was done in **Figure 13**, the average gradient of the BC line segment can be used, as indicated in **Figure 12**. Due to variations in ignition voltage and amounts of charge transferred per filament, as well as variations in the number of streamers per discharge cycle, it is advised to

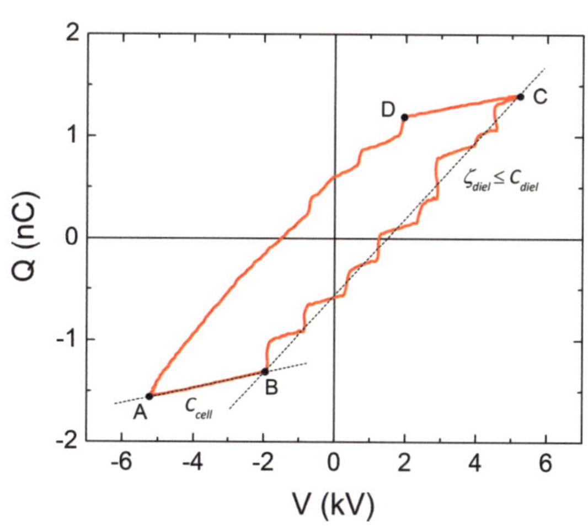

Figure 12.
Stepped Q-V diagram from a single AC cycle in a 100 kHz asymmetric planar DBD with a 3 mm² electrode area discharging in air. The steps of the plot indicate there are 8 streamers in the rising voltage part of the discharge cycle. Details of the reactor design can be found in [12].

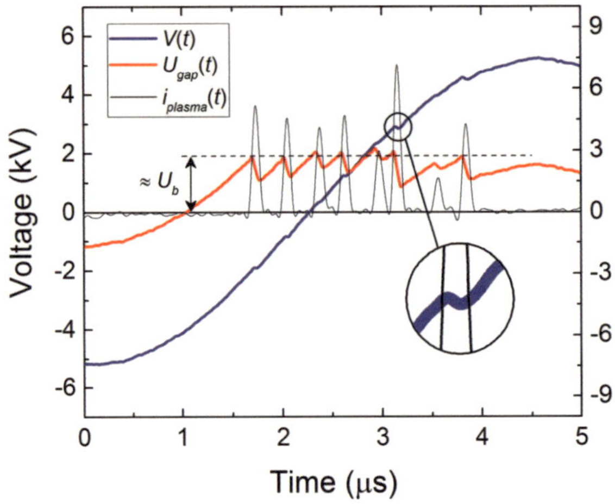

Figure 13.
Applied voltage $V(t)$*, gap voltage* $U_{gap}(t)$ *and plasma current* $i_{plasma}(t)$ *for the data in* ***Figure 12****. During a filamentary current pulse, the applied voltage* $V(t)$ *suffers a small drop due to the additional load applied to the generator, but the reactor-averaged gap voltage* $U_{gap}(t)$ *suffers a much stronger drop due to the sudden charging of the dielectric at the location of the filament. Each time* $U_{gap}(t)$ *returns to the approximate value of the burning voltage* U_b*, a new filament will ignite.*

average over a large number of applied voltage cycles when determining both ξ_{diel} and C_{diel} (see also Section 3.1) Note that once C_{diel} is determined in this way, it can subsequently be used in analysis of single applied voltage cycles as well.

Since reactors for study of individual discharges are necessarily small, there is likely to be a large parasitic reactor capacitance, which enlarges the gradient C_{cell} of the AB line segment in **Figure 12**. In these cases it is worthwhile to simply calculate an expected maximum reactor capacitance $C_{cell,calc}$ from the DBD geometry and adding the parasitic reactor capacitance $C_{parasitic} = C_{cell} - C_{cell,calc}$ as an additional parallel capacitor to the equivalent circuit of **Figure 4**, see e.g. [12, 23, 24]. A convenient expression for calculating $C_{parasitic}$ in plane-parallel configurations is:

$$C_{parasitic} = \left(1 + \frac{d_{diel}}{\varepsilon_r d_{gap}}\right) C_{cell} - \frac{d_{diel}}{\varepsilon_r d_{gap}} \xi_{diel,\,sat}, \tag{16}$$

with d_{gap}, d_{diel} and ε_r the gap width, dielectric thickness and dielectric constant, respectively, and $\xi_{diel,sat}$ the average gradient of the BC line segment at high applied voltage (at which point it saturates to a constant value, as in **Figure 6**). The results depicted for $U_{gap}(t)$ in **Figure 13**, with a nearly constant voltage U_b at each discharge ignition, was only achieved after taking into account a parasitic capacitance $C_{parasitic}$ = 45 fF on a total cell capacitance C_{cell} = 76 fF. Note also that the dielectric capacitance to be used in the equivalent circuit is $C_{diel} = \xi_{diel,sat} - C_{parasitic}$ [21].

If the DBD features only one electrode covered with a dielectric, as is the case in **Figure 12**, only one half of the discharge cycle may be stepped, while the other shows a continuous discharge during at least part of the discharging slope. The purely filamentary discharge occurs when the dielectric side acts as cathode, with filaments propagating towards the dielectric in the same direction as the current, while a Townsend-like discharge can occur when the metal electrode serves as cathode [5, 12, 52, 53]. In asymmetric cases in general (whether due to geometry or driving voltage), the average gradients of BC and DA may be somewhat different and it may be required to use a value for C_{diel} averaged over both half-cycles. Using constant values for all capacitive elements in the equivalent circuit model is a requirement to ensure continuity of charge, or more complex models must be considered instead.

4. Single filaments and filament distributions

As we saw in the previous section, it is possible to discern individual filaments in Q-V diagrams. The existence of these individual filaments in DBDs was first discovered by Buss in 1932 [54]. In his work, he used photographic glass plates as dielectrics from which a number density of 15 filaments/cycle/cm^2 could be determined. Similar photographic methods have resulted in number densities between 3 and 40 filaments/cycle/cm^2 for DBDs in air at atmospheric pressure [55–58]. However, filament footprints tend to overlap during a half-cycle and only a very limited number of cycles can be studied in one exposure, making accurate counting difficult. Monitoring the current through the DBD is, therefore, the preferred method for filament counting. Moreover, this allows the transferred charge/filament to be derived as well.

For filament counting to work, filaments need to be sufficiently spread out over time. Since this is not necessarily the case, studies of filamentary current are often restricted to geometries favouring a single filament per voltage cycle [55, 59]. A small diameter current probe embedded in a larger planar DBD has also been used, but only to obtain data on individual current pulses [57, 60, 61]. Many literature sources use relatively large electrode surface areas >1 cm^2, combined with direct digital post-processing on the DBD current $i(t)$ to determine single filament properties [62–68]. In this section, results are presented for a plane-parallel DBD with a surface area of 7 mm^2, large enough to have multiple interacting filaments per half-cycle, but small enough to prevent significant overlap between them in time. The measurement circuit in this case is an R-C circuit, with a decay constant $\tau = RC$ lower than the duration of a single half-cycle, see **Figure 14**. The voltage $V_{\mathrm{m}}(t)$ from the capacitor is passed through an analog circuit which first corrects for the capacitive displacement current by subtracting a signal $(C_{gap}/C_m)V(t)$ (see also Eq. (3)), followed by the creation of a second, delayed signal $V_{\mathrm{m}}(t+\Delta t)-(C_{gap}/C_m)V(t+\Delta t)$. By subtracting the delayed signal from the original signal (with $\Delta t \approx 50$ ns), a pulse train is created in which the height of each pulse corresponds to the horizontal charge step in a stepped Q-V diagram; see **Figure 14** for a simplified schematic of this approach. Using an ADC, robust post-processing can then be performed, in which transferred charge/filament can be determined with high fidelity, as well as

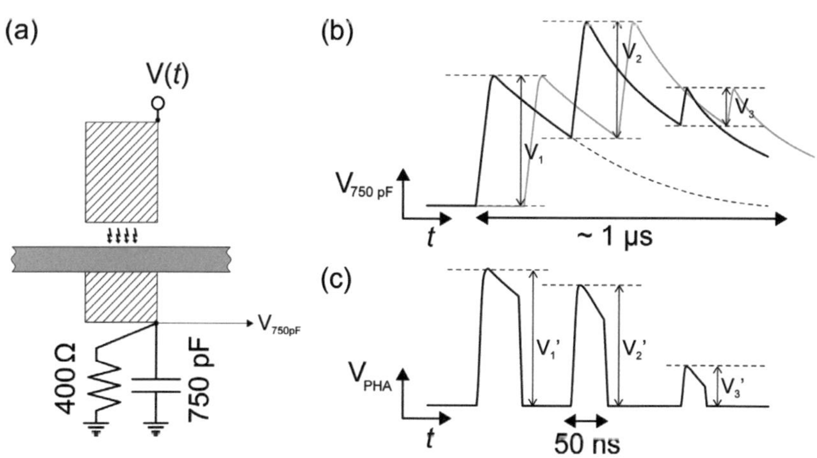

Figure 14.
Schematic representation of a pulse measurement circuit. (a) the DBD with R-C circuit, from which the voltage on a 750 pF capacitor is measured. (b) the measured voltage signal (black line) is duplicated and delayed (grey line), and (c) the two signals subtracted from one another provide a measure of the voltage step during each pulse.

allowing for identification of near-simultaneous discharges by monitoring the occurrence of multiple rising slopes within a single time interval Δt. Note that the pre-factor from Eq. (3) needs to be applied in post-processing to obtain the actual charge transferred by the plasma, instead of the transferred charge as it appears in the external circuit. A more detailed description of this pulse counting method can be found in Ref. [12].

Figure 15 shows results obtained by applying this method to a multitude of DBD configurations. **Figure 15(a)** shows the average (or mean) charge transferred per filament, while **Figure 15(b)** shows the number density of filaments occurring during a half-cycle. Several scaling laws are revealed in this data, which are irrespective of the type of dielectric used: average charge/filament increases with increasing gap width d_{gap} and decreases with dielectric thickness d_{diel}. Furthermore, the lower both d_{gap} and d_{diel}, the more filaments can be ignited per half-cycle at a given applied voltage amplitude. When using thin gaps and thin dielectrics, plasma current densities can be as high as 100 A/m^2 averaged over time and area. With filament channels in air possessing a cross-section of about 10^{-8} m^2 and an average duration of 20 ns [69, 70], this can actually amount to peak current densities of 10^9 A/m^2 at the locations of filaments. As **Figure 15(a, b)** show, the nature of the dielectric material seems to affect the discharges mainly by the capacitance they provide in the form of C_{diel}. In this regard, however, it is worth noting that none of the three materials presented here possess 'extreme' electrical properties, with $\varepsilon_r \leq 10$. Materials such as $BaTiO_3$, with $\varepsilon_r > 1000$ could present quite different behaviour [31, 48].

Aside from the behaviour of an average filament, the variation from filament to filament can be studied in detail using the same electrical measurements. **Figure 16** shows histograms of the charge/filament obtained for two different systems: in both configurations the same alumina dielectric with d_{diel} = 0.6 mm and gap width d_{gap} = 0.6 mm was used, while in **Figure 16(a)**, the plane metal electrode was replaced with a single rounded metal wire (Ø = 0.3 mm). In this way, **Figure 16(a)** shows the distribution of charge/filament for only a single filament per half-cycle, while **Figure 16(b)** shows the same distribution with up to 9 filaments per half-cycle. It is immediately obvious that a single filament produces a nearly symmetric distribution over many consecutive ignitions, while multiple filaments per half-cycle add up to a very skewed distribution. The degree of asymmetry in the distributions can be represented by the non-parametric skew S:

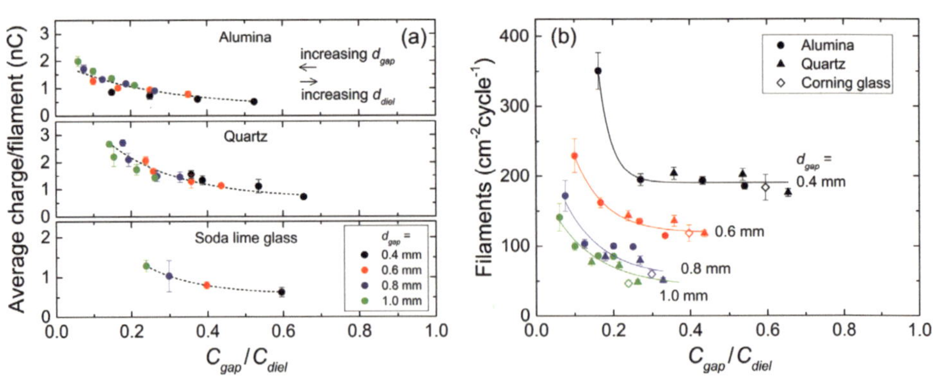

Figure 15.
*(a) Average transferred charge per filament and (b) the filament density during the positive half-cycle (*i.e. *where applied voltage is increasing) for several plane-parallel DBD configurations operated in air at 100 kHz. Only one of the electrodes is covered with a dielectric in each case. In (a) values are the average found for several applied voltage amplitudes (6–12 kV$_{pp}$ in steps of 1 kV$_{pp}$), with the error bars indicating the standard deviation. In (b) all data is for an applied voltage amplitude of 10 kV$_{pp}$ only.*

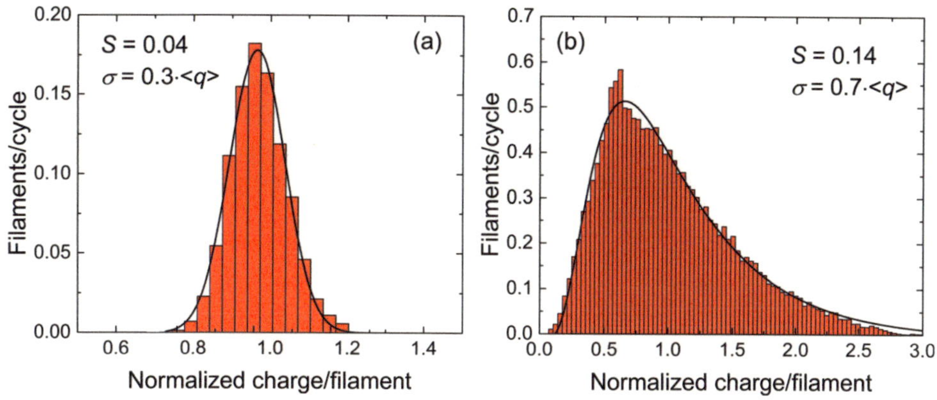

Figure 16.
Normalised charge per filament distributions per applied voltage cycle for (a) a single filament in a pin-to-plane DBD and (b) approximately 9 interacting filaments in a plane-parallel DBD with $d_{gap} = d_{diel} = 0.6$ *mm and a single alumina dielectric. The horizontal axis has been normalised to the average charge/filament <q>, which is 0.9 nC for (a) and 1.1 nC for (b). Applied voltage amplitude was 9* kV_{pp} *in both cases. The black lines are fits of (a) a normal distribution and (b) a log-normal distribution.*

$$S = \frac{\mu - \nu}{\sigma}, \tag{17}$$

with μ the mean, ν the median and σ the standard deviation. While $S \approx 0$ for a single, non-interacting filament, it increases to $S = 0.16 \pm 0.04$ for all plane-parallel DBD configurations, irrespective of applied voltage amplitude, gap width, dielectric thickness or dielectric material. Furthermore, the pin-to-plane distributions fit well to a normal distribution, while the plane-parallel configuration is described well by a log-normal distribution, as depicted in **Figure 16**. While the normal distribution for a single filament re-igniting every (half-)cycle is not surprising, multiple filaments igniting at different moments and in different places during a (half-)cycle clearly interact to produce a relatively predictable charge/filament distribution. Accounting for these log-normal-like distributions is beyond the scope of this chapter, but studies of this interaction can be found in e.g. Refs. [71,72]. The existence of this additional scaling law for plane-parallel DBDs, where charge/filament follows a log-normal distribution with $S = 0.16 \pm 0.04$, is presented here to demonstrate the potential of electrical diagnostics in assessing the behaviour of a complex, discrete system. Similar distributions could be obtained for e.g. a suitably shortened coaxial packed bed reactor, allowing computer simulations of individual filaments to be compared to real experimental data in greater detail.

5. Conclusions

In this chapter, a thorough description of charge-voltage (*Q-V*) measurement techniques is provided. Pointers are given for properly setting up *Q-V* measurements, as well as the theoretical framework required to interpret results. Causes for deviation from ideal parallelogram-shaped Q-V diagrams are treated in detail, including Q-V diagrams with variable gradients, almond-shaped and elliptical Q-V diagrams, and noisy or stepped Q-V diagrams. The chapter is rounded off with a discussion on extending *Q-V* measurements to include monitoring of individual microdischarges.

Acknowledgements

The work of Floran Peeters is part of project 'EnOp' of the Interreg V programme Flanders, Netherlands, with financial support from the European Union. This chapter is dedicated to the memory of Rein Rumphorst (1925-2018), without whom our measurements would have worked out only as well as we deserved.

Author details

Floran Peeters* and Tom Butterworth
Dutch Institute for Fundamental Energy Research (DIFFER), Eindhoven, Netherlands

*Address all correspondence to: f.j.j.peeters@differ.nl

References

[1] Aldea E, Peeters P, de Vries H, van de Sanden MCM. Atmospheric glow stabilization. Do we need pre-ionization? Surface and Coatings Technology. 2005;**200**:46-50

[2] Starostin SA, ElSabbagh MAM, Aldea E, de Vries H, Creatore M, van de Sanden MCM. Formation and expansion phases of an atmospheric pressure glow discharge in a PECVD reactor via fast ICCD imaging. IEEE Transactions on Plasma Science. 2008;**36**:968-969

[3] Starostin SA, Premkumar PA, Creatore M, van Veldhuizen EM, de Vries H, Paffen RMJ, van de Sanden MCM. On the formation mechanisms of the diffuse atmospheric pressure dielectric barrier discharge in CVD processes of thin silica-like films. Plasma Sources Science and Technology. 2009; **18**:045021

[4] Osawa N, Takashi A, Yoshioka Y, Hanaoka R. Generation of high pressure homogeneous dielectric barrier discharge in air. European Physical Journal Applied Physics. 2013;**61**:24317

[5] Osawa N, Yoshioka Y. Generation of low-frequency homogeneous dielectric barrier discharge at atmospheric pressure. IEEE Transactions on Plasma Science. 2012;**40**:2-8

[6] Golubovskii YB, Maiorov VA, Behnke JF, Tepper J, Lindmayer M. Study of the homogeneous glow-like discharge in nitrogen at atmospheric pressure. Journal of Physics D: Applied Physics. 2004;**37**(11):1346-1356

[7] Gherardi N, Massines F. Mechanisms controlling the transition from glow silent discharge to streamer discharge in nitrogen. IEEE Transactions on Plasma Science. 2001;**29**:536-544

[8] Stollenwerk L, Amiranashvili S, Boeuf J-P, Purwins HG. Formation and stabilisation of single current filaments in planar dielectric barrier discharge. European Physical Journal D: Atomic, Molecular, Optical and Plasma Physics. 2007;**44**:133-139

[9] Papageorghiou L, Panousis E, Loiseau J-F, Spyrou N, Held B. Two-dimensional modelling of a nitrogen dielectric barrier discharge (DBD) at atmospheric pressure: Filament dynamics with the dielectric barrier on the cathode. Journal of Physics D: Applied Physics. 2009;**42**:105201

[10] Wagner HE, Yurgelenas YV, Brandenburg R. The development of microdischarges of barrier discharges in N_2/O_2 mixtures—Experimental investigations and modelling. Plasma Physics and Controlled Fusion. 2005;**47**: B641-B654

[11] Peeters FJJ, Yang R, van de Sanden MCM. The relation between the production efficiency of nitrogen atoms and the electrical characteristics of a dielectric barrier discharge. Plasma Sources Science and Technology. 2015; **24**:045006

[12] Peeters FJJ. The electrical dynamics of dielectric barrier discharges. Eindhoven: Technische Universiteit Eindhoven; 2015. 197 p

[13] Banerjee AM, Billinger J, Nordheden KJ, Peeters FJJ. Conversion of CO_2 in a packed-bed dielectric barrier discharge reactor. Journal of Vacuum Science & Technology A Vacuum Surfaces an Film. 2018;**36**

[14] Kozlov KV, Wagner HE, Brandenburg R, Michel P. Spatio-temporally resolved spectroscopic diagnostics of the barrier discharge in air at atmospheric pressure. Journal of Physics D: Applied Physics. 2001;**34**: 3164-3176

[15] Sewraj N, Merbahi N, Gardou JP, Akerreta PR, Marchal F. Electric and spectroscopic analysis of a pure nitrogen mono-filamentary dielectric barrier discharge (MF-DBD) at 760 Torr. Journal of Physics D: Applied Physics. 2011;**44**:145201

[16] Chirokov A, Khot SN, Gangoli SP, Fridman A, Henderson P, Gutsol A, Dolgopolsky A. Numerical and experimental investigation of the stability of radio-frequency (RF) discharges at atmospheric pressure. Plasma Sources Science and Technology. 2009;**18**:025025

[17] Yurgelenas YV, Leeva MA. Development of a barrier discharge in air in highly nonhomogeneous electric field caused by the residual dielectric surface charges. IEEE Transactions on Plasma Science. 2009; **37**:809-815

[18] Yurgelenas YV, Wagner HE. A computational model of a barrier discharge in air at atmospheric pressure: The role of residual surface charges in microdischarge formation. Journal of Physics D: Applied Physics. 2006;**39**: 4031-4043

[19] Manley TC. The electric characteristics of the ozonator discharge. Transactions of the Electrochemical Society. 1943;**84**:83-96

[20] Pipa AV, Koskulics J, Brandenburg R, Hoder T. The simplest equivalent circuit of a pulsed dielectric barrier discharge and the determination of the gas gap charge transfer. The Review of Scientific Instruments. 2012;**83**(11): 115112

[21] Peeters FJJ, van de Sanden MCM. The influence of partial surface discharging on the electrical characterization of DBDs. Plasma Sources Science and Technology. 2014; **24**(1):015016

[22] Ashpis D, Laun M, Griebeler E. Progress toward Accurate Measurements of Power Consumption of DBD Plasma Actuators 50th AIAA Aerosp. Sci. Meet. Incl. New Horizons Forum Aerosp. Expo; 2012

[23] Falkenstein Z, Coogan JJ. Microdischarge behaviour in the silent discharge of nitrogen - oxygen and water - air mixtures. Journal of Physics D: Applied Physics. 1999;**30**:817-825

[24] Peeters FJJ, van de Sanden MCM. The influence of partial surface discharging on the electrical characterization of DBDs. Plasma Sources Science and Technology. 2015; **24**:015016

[25] Reichen P, Sonnenfeld A, von Rohr PR, Von Rohr PR. Discharge expansion in barrier discharge arrangements at low applied voltages. Plasma Sources Science and Technology. 2011;**20**: 055015

[26] Ramakers M, Michielsen I, Aerts R, Meynen V, Bogaerts A. Effect of argon or helium on the CO_2 conversion in a dielectric barrier discharge. Plasma Processes and Polymers. 2015;**12**: 755-763

[27] Mei D, Zhu X, He Y-L, Yan JD, Tu X. Plasma-assisted conversion of CO_2 in a dielectric barrier discharge reactor: Understanding the effect of packing materials. Plasma Sources Science and Technology. 2015;**24**:15011-15021

[28] Butterworth T, Allen R. Plasma-catalyst interaction studied in a single pellet DBD reactor: Dielectric constant effect on plasma dynamics. Plasma Sources Science and Technology. 2017; **26**(6):065008

[29] Snels M, Hollenstein H, Quack M. Mode selective tunneling dynamics observed by high resolution spectroscopy of the bending

fundamentals of N14 H2 D and N14 D2 H. The Journal of Chemical Physics. 2006;**125**

[30] Tu X, Verheyde B, Corthals S, Paulussen S, Sels BF. Effect of packing solid material on characteristics of helium dielectric barrier discharge at atmospheric pressure. Physics of Plasmas. 2011;**18**:1-5

[31] Butterworth T, Elder R, Allen R. Effects of particle size on CO_2 reduction and discharge characteristics in a packed bed plasma reactor. Chemical Engineering Journal. 2016;**293**:55-67

[32] Kriegseis J, Müller B, Grundmann S, Tropea C. Capacitance and power consumption quantification of dielectric barrier discharge (DBD) plasma actuators. Journal of Electrostatics. 2011; **69**:302-312

[33] Neyts EC, Bogaerts A. Understanding plasma catalysis through modelling and simulation—A review. Journal of Physics D: Applied Physics. 2014;**47**:224010

[34] Van Laer K, Bogaerts A. Improving the conversion and energy efficiency of carbon dioxide splitting in a zirconia-packed dielectric barrier discharge reactor. Energy Technology. 2015;**3**: 1038-1044

[35] Kostov KG, Honda RY, Alves LMS, Kayama ME. Characteristics of dielectric barrier discharge reactor for material treatment. 2009;**1**:322-325

[36] Kogelschatz U. Dielectric-barrier discharges: Their history, discharge physics, and industrial applications. Plasma Chemistry and Plasma Processing. 2003;**23**:1-46

[37] Wada T, Freeman GR. Temperature, density, and electric-field effects on electron mobility in nitrogen vapor. Physical Review A. 1981;**24**: 1066-1076

[38] Ryzko H. Drift velocity of electrons and ions in dry and humid air and in water vapour. Proceedings of the Physical Society. 1965;**85**:1283-1295

[39] Stano M, Safonov E, Kučera M, Matejčík Š. Ion mobility spectrometry study of negative corona discharge in oxygen/nitrogen mixtures. Chemicke Listy. 2008;**102**:1414-1417

[40] Young RA, Gatz CR, Sharpless RL, Ablow CM. New method for measuring the rates of ionic transport and loss. I. Mobility of NO+. Physics Review. 1965; **138**:A359-A370

[41] Snuggs RM, Volz DJ, Schummers JH, Martin DW, McDaniel EW. Mobilities and longitudinal diffusion coefficients of mass-identified potassium ions and positive and negative oxygen ions in oxygen. Physical Review A. 1971;**3**:477-487

[42] Kossyi IA, Kostinsky AY, Matveyev AA, Silakov VP. Kinetic scheme of the non-equilibrium discharge in nitrogen-oxygen mixtures. Plasma Sources Science and Technology. 1992;**1**:207-220

[43] Nemschokmichal S, Tschiersch R, Höft H, Wild R, Bogaczyk M, et al. Impact of volume and surface processes on the pre-ionization of dielectric barrier discharges: Advanced diagnostics and fluid modeling. European Physical Journal D: Atomic, Molecular, Optical and Plasma Physics. 2018;**72**:1-12

[44] Liu Y, Starostin SA, Peeters FJJ, van de Sanden MCM, de Vries H. Atmospheric-pressure diffuse dielectric barrier discharges in Ar-O_2 gas mixture using 200 kHz/13.56 MHz dual frequency excitation. Journal of Physics D: Applied Physics. 2018;**51**: 114002

[45] Liu Y, Peeters FJJ, Starostin SA, Sanden, Van De MCM, Vries, De HW.

Improving uniformity of atmospheric-pressure dielectric barrier discharges using dual frequency excitation. Plasma Sources Science and Technology. 2018; **27**:01LT01

[46] Lieberman MA, Lichtenberg AJ. Principles of Plasma Discharges and Materials Processing, 2nd Edition. John Wiley & Sons; 1994

[47] Biganzoli I, Barni R, Gurioli A, Pertile R, Riccardi C. Experimental investigation of Lissajous figure shapes in planar and surface dielectric barrier discharges. Journal of Physics Conference Series. 2014;**550**:012039

[48] Patil B, Peeters FJJ, van Rooij GJ, Medrano JA, Gallucci F, et al. Plasma assisted nitrogen oxide production from air: Using pulsed powered gliding arc reactor for a containerized plant. AIChE Journal. 2017;**64**(2):526-537

[49] Tu X, Gallon HJ, Twigg MV, Gorry PA, Whitehead JC. Dry reforming of methane over a Ni/Al2O3 catalyst in a coaxial dielectric barrier discharge reactor. Journal of Physics D: Applied Physics. 2011;**44**:274007

[50] Kozlov KV, Brandenburg R, Wagner HE, Morozov AM, Michel P. Investigation of the filamentary and diffuse mode of barrier discharges in N_2/O_2 mixtures at atmospheric pressure by cross-correlation spectroscopy. Journal of Physics D: Applied Physics. 2005;**38**:518-529

[51] Savitzky A, Golay MJE. Smoothing and differentiation of data by simplified least squares procedures. Analytical Chemistry. 1964;**36**:1627-1639

[52] Naudé N, Cambronne J-P, Gherardi N, Massines F. Electrical model and analysis of the transition from an atmospheric pressure townsend discharge to a filamentary discharge. Journal of Physics D: Applied Physics. 2005;**38**:530-538

[53] Osawa N, Yoshioka Y, Hanaoka R, Mochizuki Y, Kobayashi Y, Yamada Y. Generation of uniform discharge by dielectric barrier discharge device in atmospheric-pressure air. Electrical Engineering in Japan (English Transl. Denki Gakkai Ronbunshi). 2012;**180**:1-9

[54] Buss K. Die elektrodenlose Entladung nach Messung mit dem Kathodenoszillographen. Archiv fur Elektrotechnik. 1932;**26**:261-265

[55] Akishev YS, Aponin G, Balakirev A, Grushin M, Karalnik V, et al. Role of the volume and surface breakdown in a formation of microdischarges in a steady-state DBD. European Physical Journal D: Atomic, Molecular, Optical and Plasma Physics. 2011;**61**: 421-429

[56] Chirokov A, Gutsol A, Fridman A, Sieber KD, Grace JM, Robinson KS. Self-organization of microdischarges in dielectric barrier discharge plasma. IEEE Transactions on Plasma Science. 2005; **33**:300-301

[57] Braun D, Kuchler U, Pietsch GJ. Microdischarges in air-fed ozonizers. Journal of Physics D: Applied Physics. 1991;**24**:564-572

[58] Kogelschatz U, Eliasson B, Egli W. Dielectric-barrier discharges. Principle and applications. Journal de Physique IV. 1997;**7**:47-66

[59] Höft H, Kettlitz M, Hoder T, Weltmann KD, Brandenburg R. The influence of O_2 content on the spatio-temporal development of pulsed driven dielectric barrier discharges in O_2/N_2 gas mixtures. Journal of Physics D: Applied Physics. 2013;**46**:095202

[60] Eliasson B, Hirth M, Kogelschatz U. Ozone synthesis from oxygen in dielectric barrier discharges. Journal of Physics D: Applied Physics. 1987;**20**: 1421-1437

[61] Hirth M. Teilprozesse bei der Ozonerzeugung mittels stiller elektrischer Entladungen. Beitrage aus der Plasmaphys. 1980;**20**:1-14

[62] Gibalov VI, Pietsch GJ. Dynamics of dielectric barrier discharges in different arrangements. Plasma Sources Science and Technology. 2012;**21**:024010

[63] Drimal J, Kozlov KV, Gibalov VI, Samoylovich VG. On value of transferred charge in silent discharge under atmospheric pressure. Czechoslovak Journal of Physics. 1988; **38**:159-165

[64] Drimal J, Gibalov VI, Samoylovich VG. The magnitude of the transferred charge in the silent discharge in oxygen. Czechoslovak Journal of Physics. 1987; **37**:1248-1255

[65] Gibalov VI, Dřímal J, Wronski M, Samoylovich VG. Barrier discharge the transferred charge and ozone synthesis. Contributions to Plasma Physics. 1991; **31**:89-99

[66] Jidenko N, Petit M, Borra JP. Electrical characterization of microdischarges produced by dielectric barrier discharge in dry air at atmospheric pressure. Journal of Physics D: Applied Physics. 2006;**39**:281-293

[67] Ozkan A, Dufour T, Silva T, Britun N, Snyders R, et al. The influence of power and frequency on the filamentary behavior of a flowing DBD – Application to the splitting of CO2. Plasma Sources Science and Technology. 2016;**25**:025013 (11pp)

[68] Ozkan A, Dufour T, Bogaerts A, Reniers F. How do the barrier thickness and dielectric material influence the filamentary mode and CO_2 conversion in a flowing DBD? Plasma Sources Science and Technology. 2016;**25**:045016

[69] Wagner HE, Brandenburg R, Kozlov KV, Sonnenfeld A, Michel P. The barrier discharge : Basic properties and applications to surface treatment. Vacuum. 2003;**71**:417-436

[70] Braun D, Gibalov VI, Pietsch GJ. Two-dimensional modelling of the dielectric barrier discharge in air. Plasma Sources Science and Technology. 1992;**1**:166

[71] Chirokov A, Gutsol A, Fridman A, Sieber KD, Grace JM, Robinson KS. Analysis of two-dimensional microdischarge distribution in dielectric-barrier discharges. Plasma Sources Science and Technology. 2004; **13**:623-635

[72] Stollenwerk L. Interaction of current filaments in dielectric barrier discharges with relation to surface charge distributions. New Journal of Physics. 2009;**11**(10):103034

Chapter 3

Simulation on the Surface Charge Behaviors of Epoxy Insulator by Corona Discharge

Boxue Du, Hucheng Liang and Jin Li

Abstract

A majority of the high voltage (HV) electrical equipment which has solid-gas insulation has suffered greatly from the accumulation of the surface charges generated from the corona discharge. The local electric field may be distorted by the surface charge's existence and in turn causes the surface flashover faults in excessive circumstances. Consequently, it's significant to work out the mechanism of the procedure of the surface charge accumulation. A simulation model which combines both the charge trapping-detrapping procedure and the plasma hydrodynamics was created. The outcome of the simulation has agreed with the experimental results. The corona discharge intensity rises in the initial stage and then reduces as time goes by. There are various shapes of the surface potential distribution curves at various times. The central value increases quickly with time first and at last becomes saturated. Surface charges are observed in the epoxy insulator's skin layer, some of them are mobile but some are captured by traps.

Keywords: DC power transmission, epoxy insulator, corona discharge, FEM simulation, surface charge, trapping and detrapping

1. Introduction

A majority of the electrical equipment which has solid-gas insulation has suffered seriously from the existence of surface charges. Electrons and/or ions generated by the corona discharge migrate under electric force and accumulate on the insulator surface. In some cases, this may cause the local electric field distortion and even the surface flashover faults [1–3]. The surface charge distribution of GIS (gas insulated switchgear) spacer was presented in Ref. [4], which concluded that the distribution of surface charge always reaches its steady state after some time. Ref. [5] measured the surface potential decay (SPD) process of epoxy resin and found that this process takes several hours. Ref. [6] used the fluorination treatment to enhance the SPD rate of epoxy resin. Ref. [7] discussed the surface charge behaviors after various pulse application. In addition, several scholars have established the drift-diffusion equations for the purpose of describing the charge trapping-detrapping procedure within the bulk of the insulator [8–12]. Moreover, the plasma hydrodynamics models are adopted widely for the purpose of simulating the gas discharge procedure [13–16]. Nevertheless, very few people have ever tried to combine the charge trapping-detrapping procedure and the plasma hydrodynamics to simulate the surface charge accumulation procedure.

This chapter presents a needle-plane model for the purpose of studying the surface charge accumulation procedure. It is the first time that the charge trapping-detrapping procedure combines with the plasma hydrodynamics. Compared with the existing simulation models for surface charge accumulation, our model has some advantages: (1) in the ionization region, the existing models used some simplified charge transport equations to simulated the generation and transport process of charged ions. Our model is closer to the reality with many physicochemical reactions (the collision ionization etc.) in consideration; (2) in the insulator bulk, the charge trapping-detrapping process is taken into consideration; (3) in our model, the charge transport parameters (carrier mobility, carrier diffusion coefficient etc.) are obtained from solving the Boltzmann equation, which is more reliable. This chapter aims at doing some fundamental researches on the procedure of the surface charge accumulation rather than guiding the engineering application. The needle-plane model has usually been adopted to do the SPD test in many published research papers, which is of great convenience for us to compare between the outcomes of experiment and simulation. This chapter may provide some help for readers to understand the surface charge accumulation process through some simulated details which is difficult to be gained from the experimental measurements.

2. Simulation model

2.1 Geometric model

According to **Figure 1**, the needle-plane electrode system's schematic diagram is considered to be axisymmetric and therefore, it is simplified to a 2D issue. Within a lot of papers, the model often added with a mesh electrode was adopted for charging the insulators for the purpose of doing surface charge measurements. The needle electrode's radius curvature is set to 50 μm in this chapter. The grounded (GND) electrode is placed which has a thickness of 0.5 mm and a radius of 20 mm. The axial distance between the insulator upper face and the needle tip is 3.5 mm.

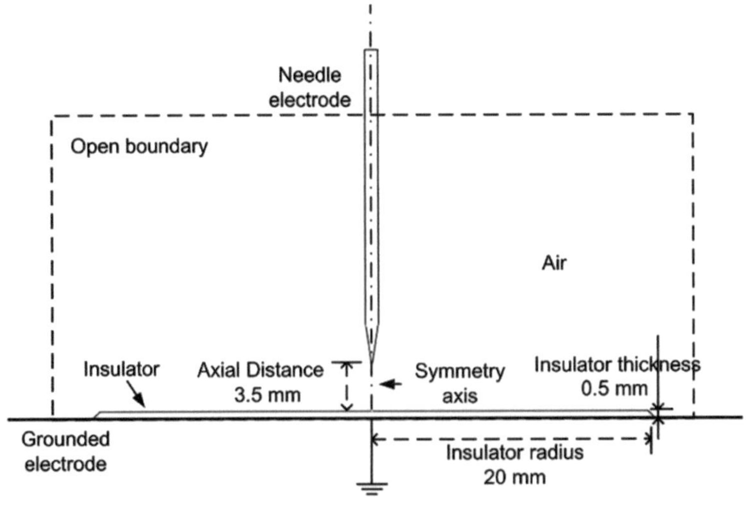

Figure 1.
A schematic diagram of the needle-plane electrode system.

2.2 Corona discharge in air

On the basis of the needle-plane model shown before, the corona discharge process is going to be simulated in this section.

The governing equations for the drift-diffusion of the electron energy density n and the electron density n_e are presented below:

$$\frac{\partial}{\partial t}(n_e)+\nabla\cdot[-n_e(\mu_e\cdot\vec{E})-D_e\cdot\nabla n_e]=R_e \tag{1}$$

$$\frac{\partial}{\partial t}(n_\varepsilon)+\nabla\cdot[-n_\varepsilon(\mu_\varepsilon\cdot\vec{E})-D_\varepsilon\cdot\nabla n_\varepsilon]+E\cdot\Gamma_e=R_\varepsilon \tag{2}$$

where the energy mobility μ_ε, the electron mobility μ_e, the energy diffusivity D_ε, and the electron diffusivity D_e are computed through working out the two-term Boltzmann equation. R_ε and R_e are the energy loss and electron source because of inelastic collisions. In this simulation, M-N three-body reactions and N two-body reactions are taken into consideration,

$$R_e=\sum_{j=1}^{N}x_jk_jN_nn_e\Delta n_{ej}+\sum_{j=N+1}^{M}x_{j1}x_{j2}k_jN_n{}^2n_e\Delta n_{ej} \tag{3}$$

$$R_\varepsilon=\sum_{j=1}^{P}x_jk_jN_nn_e\Delta\varepsilon_j \tag{4}$$

$$N_n=pT/k_B \tag{5}$$

where x_{j2} x_{j1} and x_j, are the species' mole fractions which are involved in reaction j; $\Delta\varepsilon_j$ and Δn_{ej} are the energy loss (V) and electron increment of reaction j, respectively; p is the atmospheric pressure of air (1 atm); N_n is the total neutral density (1/m^3); and k_j is the rate coefficient of reaction j (m^3/s or m^6/s), which is able to be gained through working out the two-term Boltzmann equation as well.

$$\varepsilon=\frac{n_\varepsilon}{n_e}=\frac{3}{2}T_e \tag{6}$$

where ε and T_e is the mean electron energy (V) and the electron temperature (V), respectively.

In terms of heavy species, every species' mass fraction is able to be gained through working out the equations below,

$$\rho\frac{\partial}{\partial t}(w_k)-\nabla\cdot\vec{j}_k=R_k \tag{7}$$

$$\vec{j}_k=\rho w_k\vec{V}_k \tag{8}$$

$$\vec{V}_k=D_k(\nabla w_k/w_k+\nabla M_n/M_n)-z_k\mu_k\vec{E} \tag{9}$$

where E is E-field strength (V/m); μ_k is averaged mobility of species k (m^2/(V s)); z_k is charge number of species k; M_n is s mean molar of air (kg/mol); D_k is averaged diffusion coefficient of species k (m^2/s); R_k is generation rate of species

k (kg/(m^3·s)); j_k is flux of species k; w_k is the mass fraction of species k; ρ is density of air (kg/m^3). In this simulation, *M-N* three-body reactions and *N* two-body reactions that change the species *k's* mass fraction are taken into consideration,

$$R_k = \frac{M_k}{N_A}(\sum_{j=1}^{N} k_j x_{j1} x_{j2} N_n^{\ 2} \Delta n_{kj} + \sum_{j=N+1}^{M} k_j x_{j1} x_{j2} x_{j3} N_n^{\ 3} \Delta n_{kj}) \tag{10}$$

$$x_k = \frac{M_n w_k}{M_k} \tag{11}$$

where x_{j3}, x_{j2} and x_{j1} are the species' mole fractions which are involved in reaction j; M_k is species k's molar mass (kg/mol); Δn_{kj} is species k increment of reaction j; N_A is Avogadro constant. Eq. (10) presents the relation between species k's mass fraction and the mole fraction.

Eq. (12) is the definition of the mixture averaged diffusion coefficient D_k while Eq. (13) is the definition of the mixture averaged mobility μ_k in accordance with the Relation of Einstein,

$$D_k = \frac{1 - w_k}{\sum_{j \neq k}^{Q} x_j / D_{k,j}} \tag{12}$$

$$\mu_k = \frac{eD_k}{k_B T} \tag{13}$$

where T is gas temperature (K); k_B is constant (J/K) of Boltzmann; e is unit charge (C); $D_{k,j}$ is the binary diffusion coefficient between species k and j, which is able to be evaluated through the Fuller Formula,

$$D_{k,j} = \frac{0.0101 T^{1.75} \sqrt{\frac{1}{M_k} + \frac{1}{M_j}}}{P[(\sum v_k)^{1/3} + \sum v_j)^{1/3}]^2} \tag{14}$$

where $\sum v_j$ and $\sum v_k$ is the species k and j's diffusion volume (cm^3/mol); M_j and M_k are species k and j's molar mass (kg/mol).

The boundary conditions on the surface of insulator and electrodes are defined as,

$$\vec{n} \cdot \vec{\Gamma}_e = \frac{1}{4} v_{e,th} n_e - \sum_p \gamma_p (\vec{\Gamma}_p \cdot \vec{n}) \tag{15}$$

$$\vec{n} \cdot \vec{\Gamma}_\varepsilon = \frac{5}{12} v_{e,th} n_\varepsilon - \sum_p \varepsilon_p \gamma_p (\vec{\Gamma}_p \cdot \vec{n}) \tag{16}$$

$$v_{e,th} = (\frac{8 k_B T_e}{\pi m_e})^{1/2} \tag{17}$$

$$\vec{n} \cdot \vec{j}_k = \frac{1}{4} v_{k,th} \rho w_k + \alpha \rho w_k z_k \mu_k (\vec{E} \cdot \vec{n}) \tag{18}$$

$$v_{k,th} = \left(\frac{8k_BT}{\pi m_k}\right)^{1/2} \tag{19}$$

where γ_p is the secondary emission coefficient; ε_p is the secondary electrons' mean energy (V); $\alpha{\cdot}z_k/|z_k| = 1$ if electric field can be directed towards the boundary; $\alpha{\cdot}z_k/|z_k| = 0$ if electric field can be directed away from the boundary; T is regarded as the environment temperature (K) and $v_{k,th}$ is the species k 's thermal velocity (m/s).

The definitions of the boundary conditions at the open boundary are as follows:

$$\vec{n}\cdot\nabla n_e = 0 \tag{20}$$

$$\vec{n}\cdot\nabla n_\varepsilon = 0 \tag{21}$$

$$\vec{n}\cdot\nabla w_k = 0 \tag{22}$$

Ions may become neutral species because of the surface reactions. Just several typical surface reactions will be considered for simplification in the paper.

$$N_2^+ \rightarrow N_2 \qquad O_2^+ \rightarrow O_2$$

$$O_2^- \rightarrow O_2 \qquad O_4^+ \rightarrow 2O_2$$

The plasma hydrodynamics model is made up of 19 reactions and 10 species (e.g.: O_4^+, O, O_3, O_2^-, O_2^+, N_2, N_2, O_2, N_2^+, O_2^+). Some particular physicochemical reactions are listed in **Table 1** which are considered in the model after some reduction. These collision reactions' energy losses and cross sections are extracted from

No.	Formula	Type	$\Delta\varepsilon$ (eV)	Δn_e	Rate coefficient
1	$N_2 + e \rightarrow e + N_2$	Elastic	0	0	—
2	$O_2 + e \rightarrow e + O_2$	Elastic	0	0	—
3	$N_2 + e \rightarrow 2e + N_2^+$	Ionization	−15.6	1	—
4	$O_2 + e \rightarrow 2e + O_2^+$	Ionization	−12.06	1	—
5	$N_2 + e \rightarrow e + N_2$	Excitation	−8	0	
6	$O_2 + e \rightarrow e + O_2$	Excitation	−5.5	0	—
7	$O_2^+ + O_2 + M \rightarrow O_4^+ + M$	Reaction	—	—	$2.04 \times 10^{-34}\ T^{-3.2}$
8	$O_4^+ + e \rightarrow 2O_2$	Reaction	—	−1	$1.4 \times 10^{-12}\ (300/T_e)^{0.5}$
9	$O_2^+ + e \rightarrow 2O$	Reaction	—	−1	$2.42 \times 10^{-13}\ (300/T_e)$
10	$2O_2 + e \rightarrow O_2 + O_2^-$	Reaction	—	−1	$2 \times 10^{-41}\ (300/T_e)$
11	$O_4^+ + O_2^- \rightarrow 3O_2$	Reaction	—	—	1×10^{-13}
12–13	$O_4^+ + O_2^- + M \rightarrow 3O_2 + M$	Reaction	—	—	2×10^{-37}
14–15	$O_2^+ + O_2^- + M \rightarrow 2O_2 + M$	Reaction	—	—	2×10^{-37}
16–17	$O + O_2 + M \rightarrow O_3 + M$	Reaction	—	—	2.5×10^{-46}
18	$e + N_2^+ + N_2 \rightarrow 2N_2$	Reaction	—	−1	$6.07 \times 10^{-34} T_e^{-2.5}$
19	$2e + N_2^+ \rightarrow N_2 + e$	Reaction	—	−1	$5.65 \times 10^{-27} T_e^{-0.8}$

Units: $m^3\ s^{-1}$ for two-body reactions, $m^6\ s^{-1}$ for three-body reactions, K for T_e and T; Notes: M = O_2, N_2.

Table 1.
Some typical physicochemical reactions in the corona discharge model.

papers [17, 18]. The initial electron density is set to 1×10^9 1/m^3 and the O_2 to N_2 ratio is 1:4. Δn_e means the electron increment of each reaction.

2.3 Detrapping process and charge trapping in the epoxy insulator

Electrons are going to be injected into the discharge channel when it reaches the surface of the insulator. The positive ions are converted into the neutral particles through the reactions on the surface. It can be assumed that we can also inject some holes into the insulator. Both the holes and the electrons are possible to be released from traps through thermal excitation and captured by traps when transporting. It should be noticed that the conduction on the surface is not considered in this chapter because very little tangential component exists in the electric field distribution along the surface of the insulator.

$$\frac{\partial n_{mb}}{\partial t} + \nabla \cdot \vec{J}_c^e = n_{tr} P_{de} - n_{mb} P_{tr} - R_1 n_{mb} h_{tr} - \frac{1}{\tau} \Delta n \tag{23}$$

$$\frac{\partial n_{tr}}{\partial t} = n_{mb} P_{tr} - R_2 n_{tr} h_{tr} - R_3 n_{tr} h_{mb} \tag{24}$$

$$\vec{J}_c^e = \mu_e n_{mb} \cdot \nabla V - D_e \cdot \nabla n_{mb} \tag{25}$$

$$\frac{\partial h_{mb}}{\partial t} + \nabla \cdot \vec{J}_c^h = h_{tr} P_{de} - h_{mb} P_{tr} - R_3 n_{tr} h_{mb} - \frac{1}{\tau} \Delta h \tag{26}$$

$$\frac{\partial h_{tr}}{\partial t} = h_{mb} P_{tr} - R_1 n_{mb} h_{tr} - R_2 n_{tr} h_{tr} \tag{27}$$

$$\vec{J}_c^h = -\mu_h h_{mb} \cdot \nabla V - D_h \cdot \nabla h_{mb} \tag{28}$$

where τ is the non-equilibrium carriers' lifetime (s); R_3, R_2 and R_1 are the recombination coefficients (m^3/s); P_{de} and P_{tr} are the detrapping and trapping probability of hole and electron (1/s); n_{mb} is the mobile electron density (1/m^3); μ_h and μ_e are the hole and electron's diffusion coefficient (m^2/s); V is potential (V); n_{tr} is the trapped electron density (1/m^3); J_c^h and J_c^e are the mobile electron and hole's flux (1/(m^2 s)); h_{tr} is trapped hole density (1/m^3); h_{mb} is the mobile hole density (1/m^3).

Inspired by the theories in the semiconductor physics on the non-equilibrium, the products of the hole density h_0 and mobile electron density are considered to be constants in the insulator within the condition of the thermal equilibrium.

q

$$n_0 h_0 = N_v N_c e^{-E_g / k_B T} \tag{29}$$

where E_g is the insulator's energy gap (eV); N_c and N_v are states' effective densities at conduction band bottom and the valence band top (1/m^3). Supposing the hole density and electron are h_{mb} and n_{mb} in an unbalanced condition, Δh and Δn is able to be obtained through working out Eq. (29). The procedure of the electron–hole recombination is represented by positive Δh and Δn while the electron-hole pairs' generation is represented by negative Δh and Δn.

$$n_0 h_0 = (n_{mb} - \Delta n)(h_{mb} - \Delta h) \tag{30}$$

$$\Delta n = \Delta h = \frac{(n_{mb} + h_{mb}) - \sqrt{(n_{mb} - h_{mb})^2 + 4n_0 h_o}}{2} \tag{31}$$

We can describe the boundary conditions at the insulator's upper surface as follows:

$$-\vec{n} \cdot \vec{\Gamma}_{n_{mb}} = \vec{n} \cdot \vec{j}_e \tag{32}$$

$$-\vec{n} \cdot \vec{\Gamma}_{h_{mb}} = \vec{n} \cdot \vec{j}_i \tag{33}$$

where j_i and j_e are ion and electron flux through insulator's upper surface in corona discharge model (1/m^2 s).

We can describe the boundary conditions at the insulator's lower surface as follows:

$$\vec{n} \cdot \vec{\Gamma}_{n_{mb}} = \vec{n} \cdot (-\mu_e n_{mb} \cdot \nabla V) \tag{34}$$

$$\vec{n} \cdot \vec{\Gamma}_{h_{mb}} = \vec{n} \cdot (-\mu_h h_{mb} \cdot \nabla V) \tag{35}$$

Some of the epoxy's parameters adopted in the charge trapping-detrapping model are listed in **Table 2**. Among them, some referred to papers and some were gained from experimental measurements [12, 19, 20].

2.4 Poisson equation

The p species (negative ions, positive ions and electrons) are made up by the corona discharge model and z_k is the species k's charged number. Therefore, we can describe the Poisson equation in the air as follows:

$$-\nabla^2 V = e(\sum_1^p n_k z_k) / \varepsilon_0 \tag{36}$$

Parameters	Value
Charge carrier mobility (m^2/(V s))	
μ_e	1.0×10^{-14}
μ_h	1.0×10^{-14}
Charge carrier diffusion coefficient (m^2/s)	
D_e	2.6×10^{-16}
D_h	2.6×10^{-16}
Trapping and detrapping coefficients (1/s)	
P_{tr}	7.0×10^{-3}
P_{de}	7.7×10^{-5}
Recombination coefficients (m^3/s)	
R_1	8.0×10^{-19}
R_2	8.0×10^{-19}
R_3	8.0×10^{-19}

Table 2.
Some parameters of epoxy in the trapping-detrapping model.

where species k's density is n_k (1/m^3).

We can describe Poisson equation within insulator as follows:

$$-\nabla^2 V = e(h_{mb} + h_{tr} - n_{mb} - n_{tr}) / (\varepsilon_0 \varepsilon_r) \tag{37}$$

where ε_r is the insulator's relative dielectric constant.

We add a resistor R_b for the purpose of limiting the discharge current I_p. The definition of the voltage on HV electrode is as follows:

$$V = V_0 - I_p R_b \tag{38}$$

$$I_p = -\int (\vec{n} \cdot \vec{j}_i + \vec{n} \cdot \vec{j}_e) dS \tag{39}$$

where V_0 is the voltage of power supply (V); I_p is the discharge current (A), which can be obtained by integrating the current density on the boundary of the HV electrode; V is the potential on needle electrode (V).

The open boundary is as follows:

$$\vec{n} \cdot \nabla V = 0 \tag{40}$$

2.5. Surface potential measurement

Some surface potential measurement tests are implemented for the purpose of verifying this simulation's validity. The measurement system of the needle-plane electrode surface potential is shown in **Figure 2**. The HV electrode was put 3.5 mm over the central insulator and supplied HVDC voltages on it through a power source with high voltage. Needle electrode's radius curvature was ~50 μm. The design of the insulator was like a disc of 0.5 mm thick. The insulator was slid to the position 3 mm on a rail below the probe after charging for a minute, which was adopted for measuring the distribution of the surface potential. It is not difficult to obtain the distribution of the surface potential along the radial distance through the measurement of 5 points' values from the center to the insulator's edge.

3. Results and discussion

With DC −5 kV applied on the needle, the electric potential and force line distribution are simulated and presented in **Figure 3**. The electric potential's variation is represented by the gradient color which is presented in the color legend on

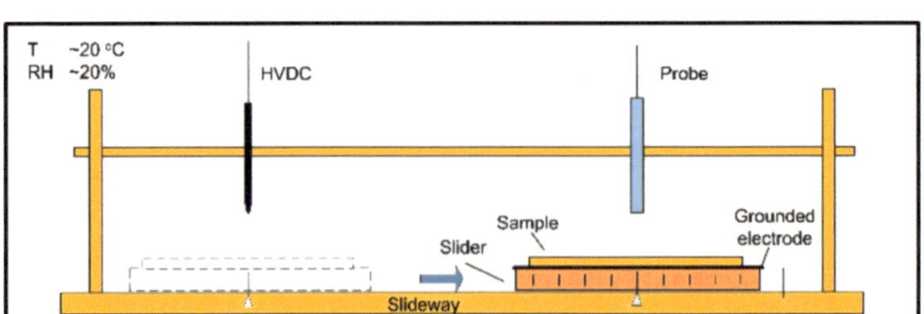

Figure 2.
A schematic diagram of the surface potential measurement system.

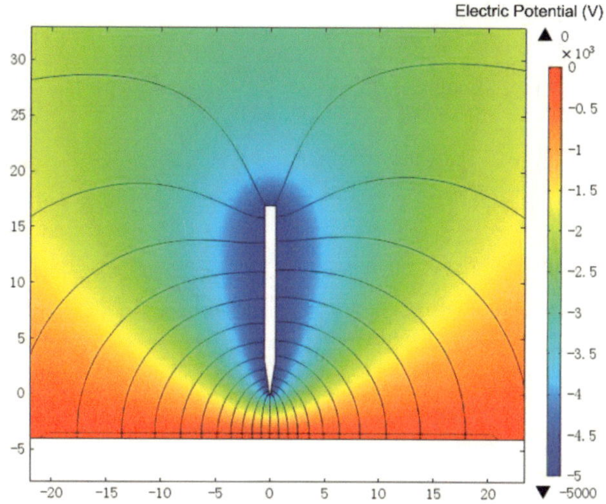

Figure 3.
The simulated distribution of electric potential and force lines.

the figure's right side. The electric force line distribution can be obtained through the calculation of the potential gradient. It can be seen that all the force lines go from the needle to the ground and stronger field strength at the needle tip is indicated by denser force lines. Consequently, the corona discharge is produced like an avalanche. Electrons with negative charge transport from the needle tip to the insulator's upper surface under electric force but ions with positive charge go oppositely, which causes the negative charges to accumulate on surface of insulator. In addition, very little tangential component of E-field exists on the surface of the insulator. Therefore, this chapter will not consider about the surface conduction.

Electron density's variation is shown in **Figure 4**. The electron density's variation is represented by the gradient color which is presented in the color legend on figures' right side. Here the E-field strength is high enough to produce electron-ion pairs and ionize the air due to the needle tip's small curvature radius. According to **Figure 4**, it can be seen that the electron avalanche's small crescent appears near the needle tip. Electrons gain higher speed from the needle tip under electric force, which produces more electrons through impact ionizations. We can see in **Figure 5b** that a clear increment exists in the electron avalanche's density and size. The electron avalanche's size becomes larger and larger at the time of moving to the insulator surface. The electron density's distribution can be shown in **Figure 4c**. The electron avalanche's head has reached the surface of the insulator at that time, which forms a clear discharge channel. Many electrons start accumulating on the surface of the insulator, which causes the E-field strength's disadvantages and the surface potential's rise. The discharge channel is barely able to be recognized from **Figure 4d**, which indicates the end of the corona discharge and surface charge accumulation.

According to **Figure 5**, the ion and electron densities are shown along the symmetry axis, which echoes **Figure 4** which is presented before. The axial coordinate $z = -3.5$ mm is the insulator surface and $z = 0$ mm is the needle tip. **Figure 5a** shows that the corona discharge is started from the needle tip. Positive charges concentrate at the tail and negative charges concentrate at the electron avalanche's head. Electrons are accelerated away with time, which produces more ions and electrons. **Figure 5b** shows that the peaks of the ion and electron densities have reached approximately 1×10^{15} $1/m^3$. It is clear that the electron avalanche grows

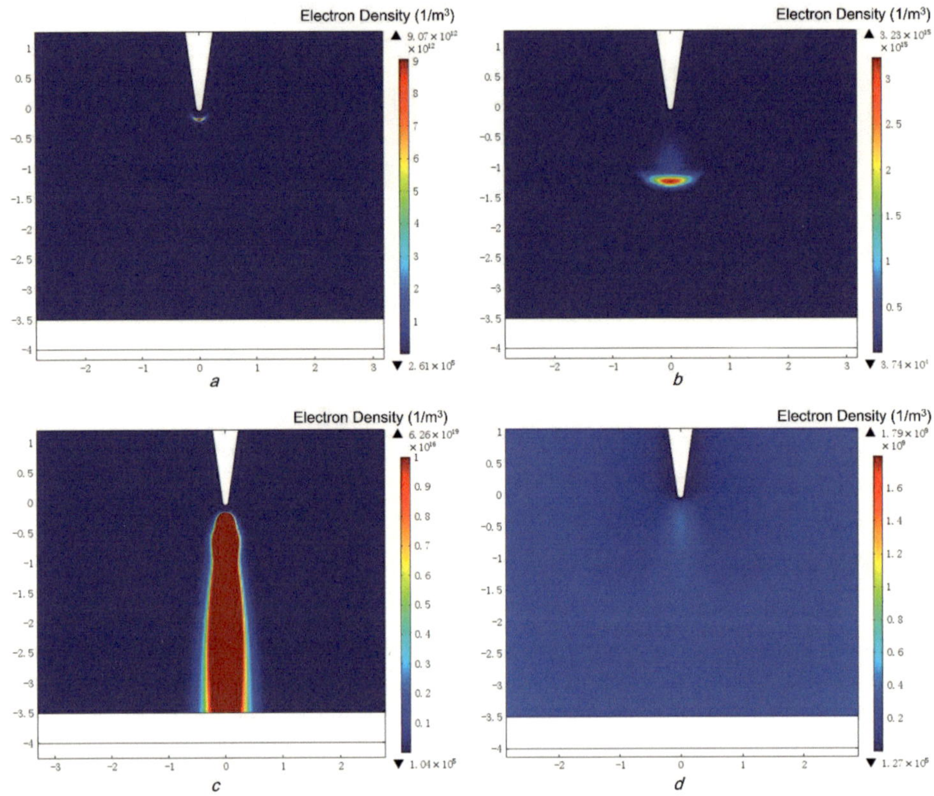

Figure 4.
The variation of electron density with the development of corona discharge: (a) 1×10^{-9} s, (b) 1.8×10^{-8} s, (c) 4.6×10^{-7} s, (d) 60 s.

sharply. The mobility of ions is much smaller than that of electrons. Therefore, they seldom move under electric force which results in the electron avalanche tail's extension. **Figure 5c** shows that the electron avalanche has reached the surface of the insulator for the purpose of forming a discharge channel. Both the ion and electron densities are increased by more than magnitude's four orders. Many negative charges have been accumulated as the discharge time goes by, which causes the field strength's disadvantages and increases the surface potential. The procedure of the corona discharge becomes weak. Thus, both the ion and electron densities reduces with time, which is presented in **Figure 5d**.

According to **Figure 6a**, the surface potential distributions are shown with the corona discharge's development. More electrons are able to be accumulated on the center of the surface because the needle electrode is above the center of the insulator directly. The peak of the surface potential appears in the central part and decreases to the edge from the central part. Additionally, the surface potential distributions' curves have various kinds of shapes at various times. The surface potential is ~2000 V at the center when time t = 1 s, which is much higher compared with that at the edge. One second is not very long that although many electrons have reached the surface of the insulator, hardly any electrons have enough time to diffuse. Consequently, the surface potential increases just by concentrating on the central surface's small zone. When time t = 30 s, it is obvious to see not only a sharp reduction but also an increment in the center of the surface potential. It's because an increasing number of surface charges have diffused with time from the central area. The surface potential gradient's decrease is even clearer when time t = 60 s but the central surface potential's growth becomes negligible.

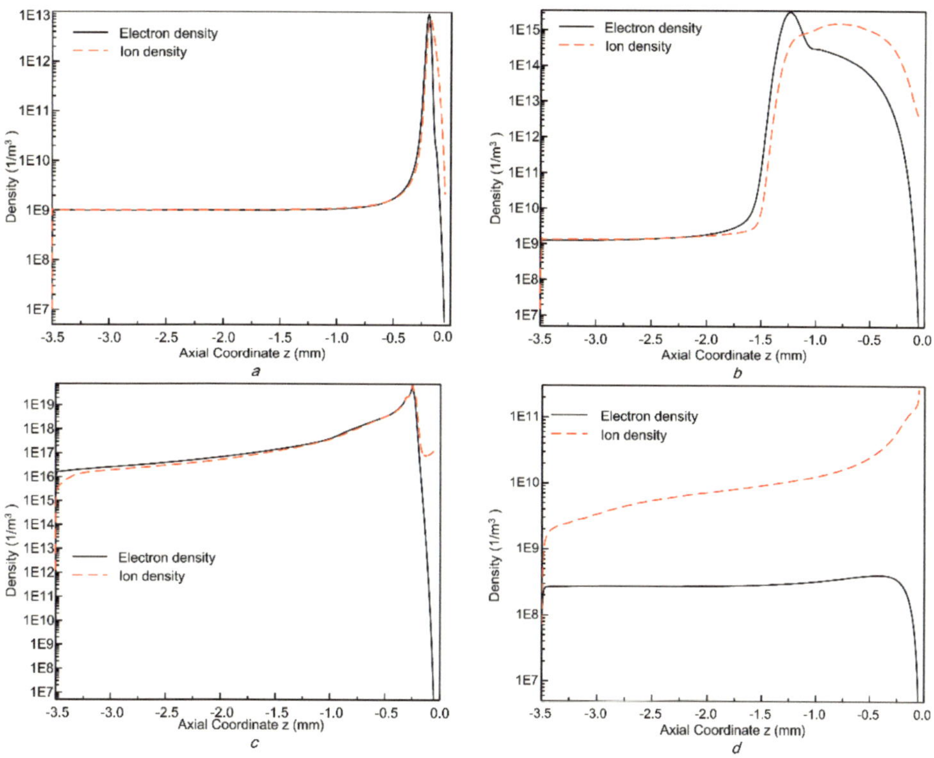

Figure 5.
The distribution of electron and ion density along the symmetry axis: (a) 1×10^{-9} s, (b) 1.8×10^{-8} s, (c) 4.6×10^{-7} s, (d) 60 s.

According to **Figure 6b**, the E-field strength distributions are shown along the symmetry axis. It's obvious that the field strength's sharp decrease occurs at the interface of insulator and air because of the epoxy insulator's higher relative dielectric constant compared with that of air. With the development of the corona discharge, an increasing number of electrons will exist on the insulator surface. The inside E-field strength's increment in insulator is caused by higher surface potential. The E-field strength at needle tip decreases as time goes by due to the decrease of the potential difference between the insulator surface and the needle tip. The E-field strength at needle tip is presented in **Figure 6b**'s margin for the purpose of seeing the outcome obviously.

When the epoxy insulator is injected with electrons, they tend to transport to GND electrode across the bulk under the built-in electric area. Several mobile electrons are possible to be captures by traps at the time of transporting and later on gain de-trapped by thermal excitation. Owing to the exceeding existence and low mobility of trapped electrons, the injected electrons are able to be observed just in insulator's skin layer. According to **Figure 7**, the distribution of trapped electrons and the mobile along symmetry axis is shown from GND electrode to insular upper surface. d = 0.5 mm is GND electrode while d = 0 mm is epoxy insulator's upper surface. According to **Figure 7a**, it is quite obvious to see that mobile electron density decreases quickly from upper surface to ground. Mobile electron increases a bit deeper in the skin layer and decreases at the upper surface. **Figure 6a** concludes that the surface potential's quick rise occurs at the beginning 1 s. In other words, most of the electrons will not be injected into epoxy insulator within the rest 59 s. Nevertheless, many electrons have gone deeper in skin layer under electric force. Consequently, electron density reduces at upper surface as time goes by. According

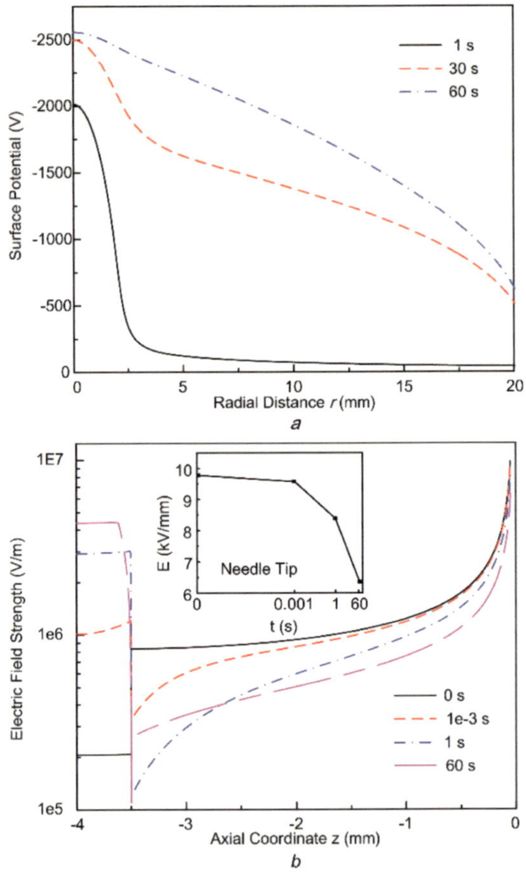

Figure 6.
The surface potential and electric field distributions: (a) surface potential and (b) electric field.

to **Figure 7b**, the trapped electrons' density increases obviously with time in the skin layer because an increasing number of mobile electrons get trapped but seldom escape at the time of transporting to the ground.

According to **Figure 8**, the central surface potential grows with time when various kinds of charging voltages are applied. It is clear central surface potential's absolute value goes up quickly. The surface potential increases sharply at the beginning 1 s under −5 kV. Later on, the growth gradually slows down with time. At last, the surface potential may reach saturation with the dissipation procedure in dynamic equilibrium and the surface charge accumulation when there is enough discharge time. When the charging voltage reduces, the surface potential in the center will spend more time to reach saturation.

According to **Figure 9**, the comparison between experimental and computational surface potential distributions is presented under −5 kV. The difference is very clear between the outcomes of the simulation and the experiment resources when t = 30 s which is presented in **Figure 9a**. The simulated value is higher than the experimental surface potential at the center. However, the experimental potential will become much higher within the area around ~ 15 mm away from the center. Firstly, it is possible that the probe's precision will not be high enough. Secondly, several seconds were taken to take off the power source and move the insulator before measuring its surface potential. Many accumulated electrons have spread in the surrounding areas from the center in the time delay, which increases the surface potential in the surrounding regions but decreases the surface potential at the

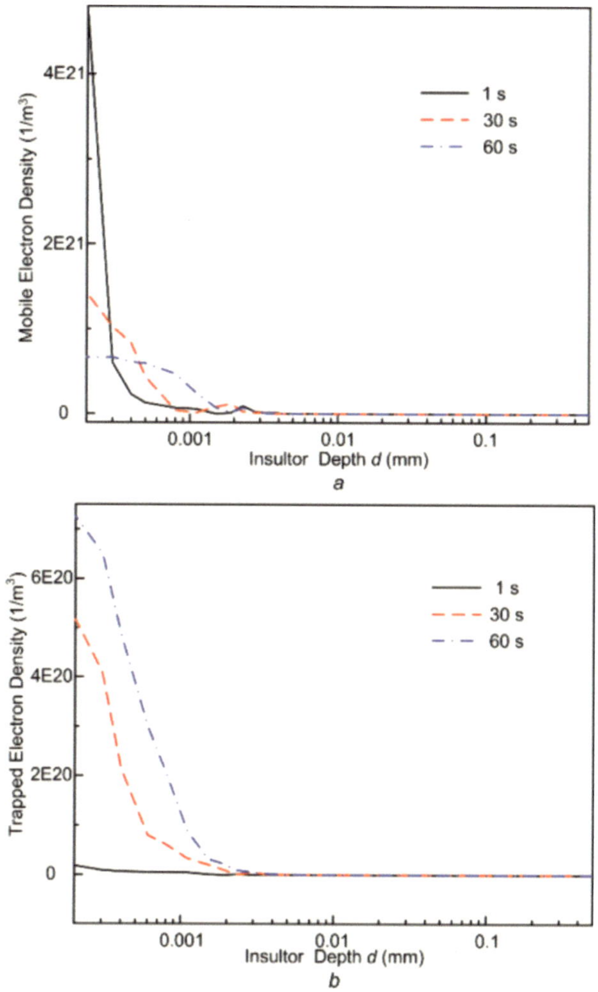

Figure 7.
The mobile and trapped electron density distribution along the symmetry axis of insulator bulk: (a) mobile electron and (b) trapped electron.

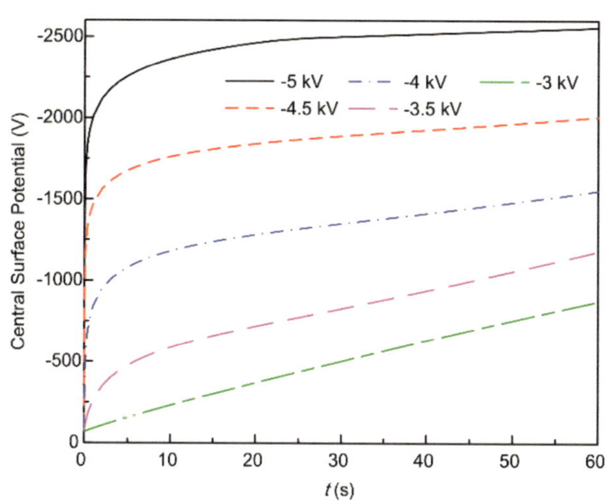

Figure 8.
The growth of central surface potential with time under different charging voltages.

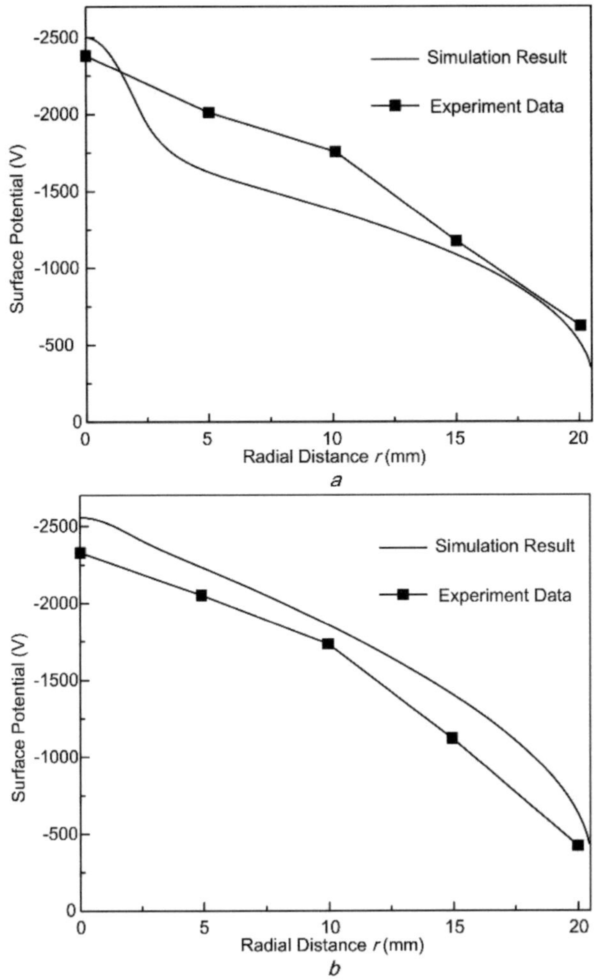

Figure 9.
The comparison between the computational and experimental surface potential distributions: (a) t = 30 s, (b) t = 60 s.

center. According to **Figure 9b**, two similar curves of experimental and computational surface potential distributions are shown when t = 60 s. nevertheless, it is evident that the simulated values are higher than the experimental resources. The time delay is possible to be responsible before the surface potential measurement.

According to **Figure 10**, epoxy insulator's central surface potentials are presented under various kinds of voltages after the charging for 60 s. It is obvious to see both the simulated and measured potential values grow up nearly in a linear way with the charging voltage at the insulator center. According to **Figure 10b**, the central potential grows with time under −5 kV. Experimental resources present a nice agreement with the outcomes of the simulation that central potential goes up quickly at the beginning seconds and at last reaches a fixed condition. Because of the diffusion of charges from center, the outcomes of the simulation are often higher than the measured surface potentials, particularly at the beginning 10 s. According to what has been discussed in **Figure 5**, the gradient of the surface potential distribution is much clearer along the radial distance at the beginning 10 s. Therefore, the central surface charges' spread around is clearer to cause a greater effect on measured surface potential. Thus, the difference between the outcomes of the simulation and the experimental resources is wider at the beginning 10 s.

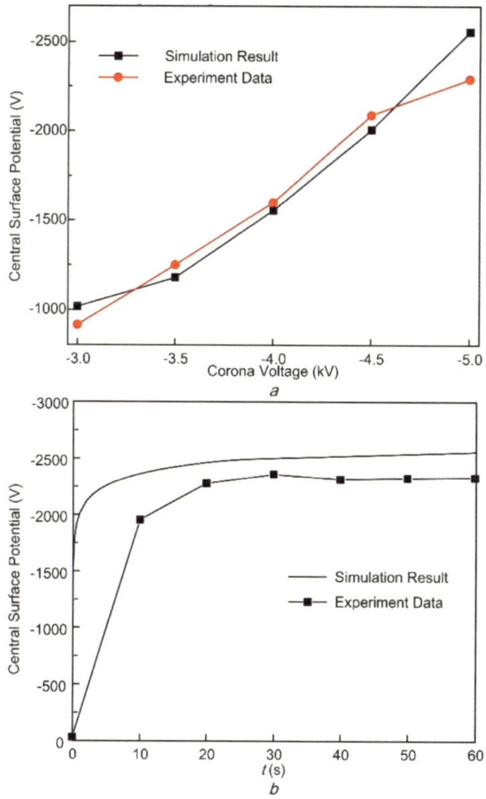

Figure 10.
The comparison between the computational and experimental central surface potentials: (a) the central surface potentials after 60s' charging and (b) the growth of central surface potential with time.

4. Conclusions

The accumulation procedure of the surface charge under the needle-plane corona discharge has been explored in This chapter and some comparisons between experimental resources and the simulation outcomes have also been made. The major conclusion is shown below:

1. In the accumulation procedure of the surface charge, the corona discharge intensity increases at the beginning and later reduces as time goes by. The epoxy insulator's surface potential has increased sharply first and later slowly reaches saturation.

2. The surface potential distributions' curves have various kinds of shapes at various periods. As time goes by, the surface potential gradient along epoxy insulator's radial distance reduces because of the surface charge's diffusion.

3. Meanwhile, a higher central surface potential exists in the epoxy insulator under higher charging voltage when it is being charged. Then, the central surface potential takes a shorter time to reach saturation.

4. The epoxy insulator's skin layer has a surface charge. These mobile electrons have a tendency to cross the insulator to the GND electrode. Among them, some get trapped and gradually get detrapped by thermal excitation.

Author details

Boxue Du, Hucheng Liang and Jin Li*
Key Laboratory of Smart Grid of Education Ministry, School of Electrical and Information Engineering, Tianjin University, Tianjin, China

*Address all correspondence to: lijin@tju.edu.cn

References

[1] Kumara S, Alam S, Hoque IR, et al. DC flashover characteristics of a polymeric insulator in presence of surface charges. IEEE Transactions on Dielectrics and Electrical Insulation. 2012;**19**(3):1084-1090

[2] Kumara S, Serdyuk YV, Gubanski SM. Simulation of surface charge effect on impulse flashover characteristics of outdoor polymeric insulators. IEEE Transactions on Dielectrics and Electrical Insulation. 2010;**17**(6): 1754-1763

[3] Du BX, Li ZL. Surface charge and DC flashover characteristics of direct-fluorinated SiR/SiO_2 nanocomposites. IEEE Transactions on Dielectrics and Electrical Insulation. 2015;**21**(6): 2602-2610

[4] Nakanishi K, Yoshioka A, Arahata Y, et al. Surface charging on epoxy spacer at DC stress in compressed SF6 gas. IEEE Transactions on Power Apparatus and Systems. 1983;**102**(12):3919-3927

[5] Sato S, Zaengl WS, Knecht A. A numerical analysis of accumulated surface charge on dC epoxy resin spaces. IEEE Transactions on Electrical Insulation. 1987;**22**(3):333-340

[6] Mohamad A, Chen G, Zhang Y, et al. Mechanisms for surface potential decay on fluorinated epoxy in high voltage DC applications. IEEE Conference on Electrical Insulation and Dielectric Phenomena (CEIDP). 2014. pp. 863-866

[7] Du BX, Zhang JW, Gao Y. Effect of nanosecond rise time of pulse voltage on the surface charge of epoxy/TiO_2 nanocomposites. IEEE Transactions on Dielectrics and Electrical Insulation. 2013;**20**(20):321-328

[8] Roy SL, Segur P, Teyssedre G, Laurent C. Description of bipolar charge transport in polyethylene using a fluid model with a constant mobility: Model prediction. Journal of Physics D: Applied Physics. 2003;**37**(2):298

[9] Min D, Li S. Simulation on the influence of bipolar charge injection and trapping on surface potential decay of polyethylene. IEEE Transactions on Dielectrics and Electrical Insulation. 2014;**21**(4):1627-1636

[10] Chen G. A new model for surface potential decay of corona-charged polymers. Journal of Applied Physics. 2010;**43**(5):55405-55411

[11] Zhou F, Li J, Yan Z, et al. Investigation of charge trapping and detrapping dynamics in LDPE, HDPE and XLPE. IEEE Transactions on Dielectrics and Electrical Insulation. 2017;**23**(6):3742-3751

[12] Min D, Li S, Li G. The effect of charge recombination on surface potential decay crossover characteristics of LDPE. IEEE International Symposium on Electrical Insulating Materials (ISEIM). 2014. pp. 104-107

[13] He W, Liu XH, Yang F, et al. Numerical simulation of direct current glow discharge in air with experimental validation. Journal of Applied Physics. 2012;**51**(51):6001

[14] Nahomy J, Ferreira CM, Gordiets B, et al. Experimental and theoretical investigation of a N_2-O_2 DC flowing glow discharge. Journal of Physics D. 1995;**28**(4):738

[15] Choi J, Matsuo K, Yoshida H, et al. Characteristics of a DC-driven atmospheric pressure air microplasma jet. Journal of Applied Physics. 2008; **47**(8):6459-6463

[16] Tochikubo F, Arai H. Numerical simulation of streamer propagation and

radical reactions in positive corona discharge in N_2/NO and N_2/O_2/NO. Journal of Applied Physics. 2002;**41** (2A):844-852

[17] Itikawa Y, Ichimura A. Cross sections for collisions of electrons and photons with atomic oxygen. Journal of Physical and Chemical Reference Data. 1990;**19**(3):637-651

[18] Itikawa Y. Cross sections for electron collisions with nitrogen molecules. Journal of Physical and Chemical Reference Data. 2006;**34**(38): 1-20

[19] Zhou T, Chen G, Liao R, Xu Z. Charge trapping and detrapping in polymeric materials: Trapping parameters. Journal of Applied Physics. 2009;**106**(12):644-637

[20] Chen G, Zhao J, Zhuang Y. Numerical modeling of surface potential decay of corona charged polymeric material. IEEE International Conference Solid Dielectrics (ICSD). 2010. pp. 1-4

Section 2

Plasma Applications

Chapter 4

Applications of Dielectric Barrier Discharge Microplasma

Kazuo Shimizu, Jaroslav Kristof and Marius Gabriel Blajan

Abstract

Dielectric barrier discharge microplasma is a nonthermal plasma discharge at atmospheric pressure which due to the micrometer size dielectric layer between the grounded and high-voltage energized electrodes enables to drive the device at less than 1 kV. Microplasma is an economical and ecological alternative for conventional technologies used for NOx removal, indoor air cleaning, surface treatment of polymers, biomedical applications such as transdermal drug delivery, or as an actuator. In this chapter, microplasma applications such as indoor air purification, skin treatment for drug delivery, particle removal, and flow control are presented.

Keywords: dielectric barrier discharge, microplasma, indoor air purification, plasma drug delivery, particle removal, electrohydrodynamic flow, plasma actuator

1. Introduction

Microplasma is a term used typically for referring to gas discharges that have dimensions ranging from few micrometers up to few millimeters. The breakdown voltage which is the voltage that is required to ignite a discharge depends on the product of pressure p and discharge gap d, that is also known as the Paschen curve. According to the Paschen curve, at atmospheric pressure, the breakdown voltage can be kept at low values if the discharge gap is below 1 mm. Thus the typical operating parameters of microplasmas (pressures up to and exceeding 1 atmosphere and discharge gaps below 1 mm) correspond to the values of p and d product similar to the values of the large-volume low-pressure plasmas but with much higher energy densities [1–3]. The microplasma presented in this chapter is a nonthermal plasma discharge at atmospheric pressure. It is a type of dielectric barrier discharge (DBD) generated using a configuration having a dielectric material covering the electrodes and a narrow discharge gap of micrometer size between electrodes. Because it is generated at relatively low discharge voltages of about 1 kV, the reactor has small dimensions and requires only small-sized power supplies. Considering the electrodes covered by a dielectric layer with a dielectric constant of $\varepsilon_r = 10^4$ and according to the law that states the electric displacement field inside the capacitor formed by the two dielectric layers and the air gap is constant, it results in the air gap an electric field 10^4 times higher than the electric field in the dielectric layers. At a discharge gap of 10 μm, it has obtained a high electric field $E = 10^7$–10^8 V/m that assures the formation of nonthermal plasma at a discharge voltage about 1 kV. The temperature estimation of the microplasma discharge using emission spectroscopy shows high electron and low rotational temperatures. These temperatures are specific for nonequilibrium plasmas [4].

Despite the small size, the development still continues in order to improve the technology and further miniaturization.

As a technology for indoor air purification [5–7], microplasma was used to remove hexadecane ($C_{16}H_{34}$) which is a semivolatile organic compound (SVOC) that has its origin from diesel oil [8–11]. Due to the exhaust gases present in the atmosphere, the hexadecane also enters indoors. The concentration of hexadecane decreased from 3000 to 105 ppm after 120 minutes of microplasma processing.

In the field of transdermal drug delivery and cosmetics, microplasma can change the surface properties of the skin and increase adsorption of relevant substances that are important for creams, detergents, and other formulations applied on the skin surface [12–15]. Moreover, changes in the lipid barrier of the skin induced by microplasma can make it easier to incorporate molecules in the skin or through the skin.

Microplasma can be used as a technology for particle removal. The removal or transfer of fine particles is important for global environmental problems such as PM 2.5, semiconductor industry dust contamination, [16] or for the exploration activities on the moon [17, 18]. Due to the electrostatic and dielectrophoretic forces, particles can be removed from surfaces of solar cells or from components in electronic industry in the case of dust contamination [19–21].

Microplasma actuator is a technology that can replace mechanical actuators. Due to the lack of moving parts and fast response, it has advantages over the conventional actuators [22–24]. The electrohydrodynamic (EHD) phenomenon, which occurs due to the momentum transfer from ions accelerated by an electric field to neutral molecules by collision, generates flow. A multielectrode geometry of the microplasma actuator and the possibility to drive various parts of the actuator independently using FET switches were used; thus the flow direction was controlled. The flow characteristics at various time intervals were measured by a high-speed camera. The small dimension of microplasma actuator and the light emission near the active electrodes make the experimental investigation near the electrodes difficult. Numerical simulation software based on the Suzen-Huang model coupled with Navier-Stokes equations was developed; thus additional information about the flow characteristics was obtained near the electrodes due to numerical simulation [25–27].

With these applications, microplasma technology has proven to be a reliable alternative for a wide range of conventional technologies with advantages of being ecological, economical, and easy to integrate with other systems.

2. Applications of dielectric barrier discharge microplasma

Various applications of dielectric barrier discharge microplasma make this technology a versatile and effective tool used for indoor air purification, transdermal drug delivery, particle removal, or flow control.

2.1 Indoor air purification

Indoor air quality (IAQ) is an important factor for a healthy life because people spend a major part of their time inside buildings. Among the factors that worsen the IAQ are smells, fine particles, and various chemical substances [5–7]. Technologies for purifying the indoor air are studied, and equipment is already on the market. Plasma is a technology that is applied for air purification. The main nonthermal technologies for air purification could be roughly classified in methods using dielectric barrier discharge [28, 29] and corona discharge [30, 31]. Each method has advantages, but the dielectric barrier discharge is more suitable to an easy integration in other systems. Moreover dielectric barrier discharge microplasma has a discharge gap of micrometer order and requires a discharge voltage of about 1 kV; thus only a small-sized power supply is needed. Various active species, ozone (O_3), and ultraviolet light (UV) are generated by plasma discharge in air at atmospheric pressure [8, 9]; thus chemical substances can be decomposed or bacteria sterilized [10, 11]. One of the substances that are studied to be decomposed by microplasma is hexadecane ($C_{16}H_{34}$) [32–34]. Hexadecane is a semivolatile organic compound (SVOC), and because it is present in diesel oil, it enters inside buildings from outside. A bad smell is caused by hexadecane and also affects the respiratory organs. It is a large-molecular-weight substance and thus not easy to decompose because of many carbon–carbon bonds that are strong and have low chemical reaction.

The results of the hexadecane removal process by microplasma were investigated by using the Fourier transform infrared spectroscopy (FTIR) (SHIMADZU, Kyoto, Japan, FTIR-8400S), and a gas chromatograph-mass spectrometer (GC-MS) (SHIMADZU, GCMS-QP2010). For the GC-MS measurements, nonpolar columns (Restek, Bellefonte, PA, USA, Rxi-1ms fused silica capillary column (30 m × 0.25 mm ID, 0.25 μm film thickness)) consisting of Crossbond (dimethyl polysiloxane) were used.

2.1.1 Microplasma electrode for indoor air purification

The microplasma electrode used for the hexadecane removal process is shown in **Figure 1**. Two metallic electrodes with 1 mm thickness, perforated with holes (φ = 5 mm, aperture ratio: 41.5%), are coated with dielectric. A 30-μm gap is obtained in between the electrodes and using a resin at periphery to fix them. Plasma was generated around the holes and in between the electrodes while the gas flowed through the holes and was processed by microplasma discharge [35–37]. The microplasma discharge started when applying a sinusoidal voltage with amplitude 400 V0 - p (0 to peak) and frequency 27 kHz in air using a neon transformer.

Applied voltage was controlled using a variable autotransformer.

The waveforms of voltage and corresponding current when the electrode was energized at 550 V0 − p in air are shown in **Figure 2**. The spikes convoluted in the current waveform appeared due to plasma generation.

2.1.2 Hexadecane decomposition

The following process was used for hexadecane removal using the microplasma electrode:

1. The microplasma electrode was placed inside the small chamber (11.4 L), and the chamber was closed.

2. About 3000 ppm of hexadecane was vaporized in the chamber.

3. The voltage applied to the microplasma electrodes for 120 minutes was 550 V0 − p.

Pure air (CO < 1 ppm, $CO_2 < 1$ ppm, THC < 1 ppm) from the gas cylinder was used. The amount of hexadecane before and after microplasma treatment was measured by using FTIR. The spectrum before and after vaporizing 3000 ppm of hexadecane in the chamber filled with room air (40–50%RH) is shown in **Figure 3**. In the upper part of the figure, using arrows, the alkanes or alkenes which were generated during the vaporization process of hexadecane are shown.

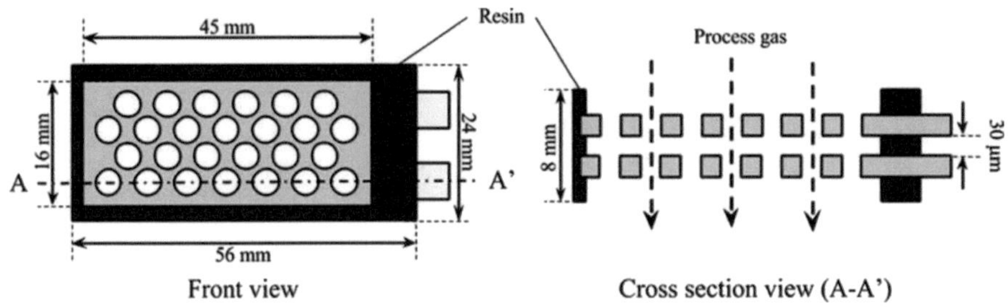

Figure 1.
Microplasma electrode.

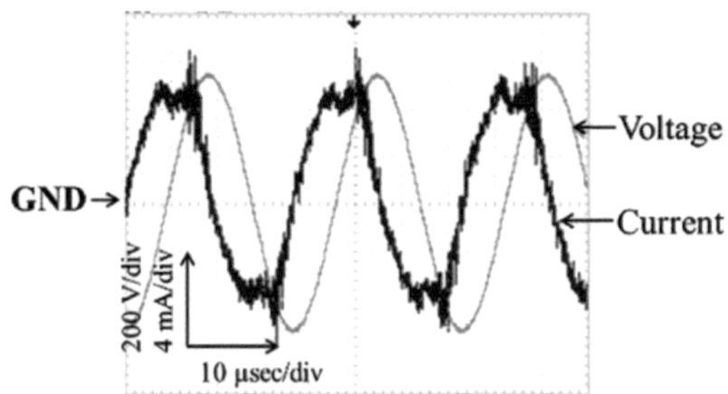

Figure 2.
Waveforms of discharge voltage and corresponding discharge current.

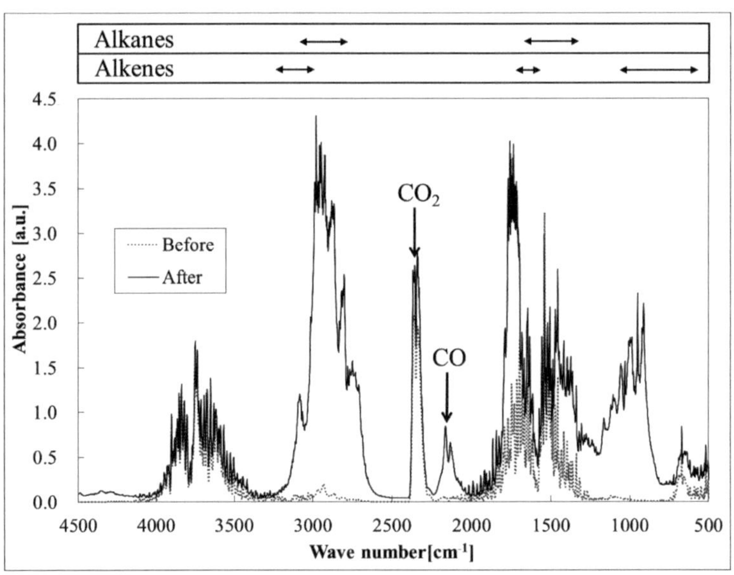

Figure 3.
Spectrum before and after vaporizing of hexadecane.

In **Figure 4**, the FTIR spectrum obtained before and after 120 minutes without using microplasma in the chamber filled with room air and containing 3000 ppm of hexadecane is shown. The spectrum after 120 minutes is the "natural decay." Even without the microplasma, hexadecane decreased after vaporization due to the condensation on the chamber walls. In **Figure 5**, the spectrum before and after 120 minutes of plasma processing in the chamber filled with room air and an initial concentration of 3000 ppm of hexadecane is shown.

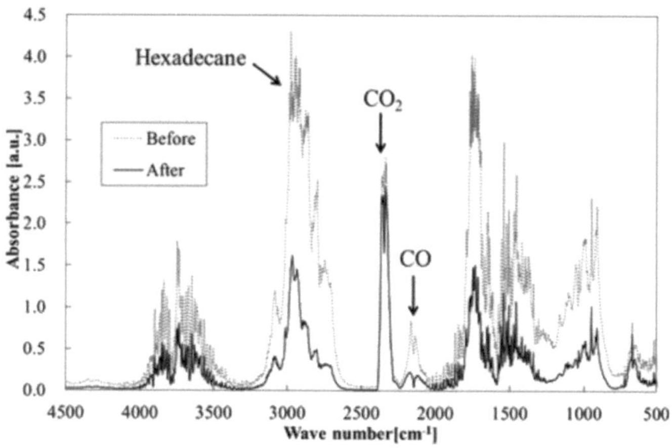

Figure 4.
Spectrum before and after 120 minutes without the plasma processing.

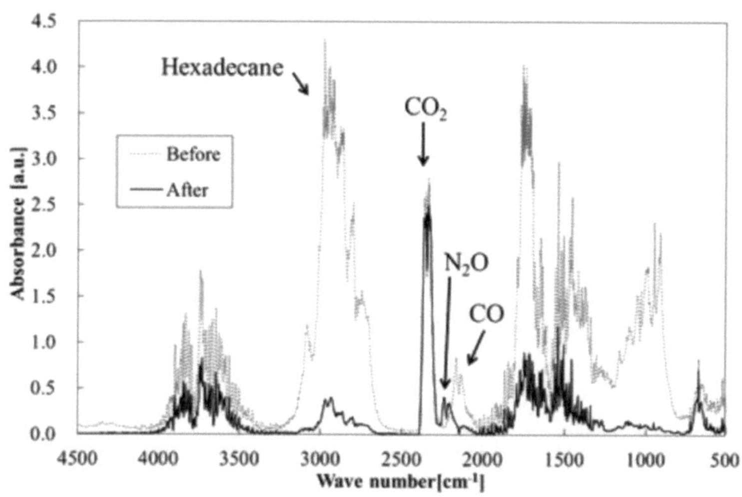

Figure 5.
Spectrum before and after 120 minutes of plasma processing.

The peak of vibrational band corresponding to hexadecane and CO decreased after microplasma treatment and that of N_2O increased. The FTIR measurement of other bands was difficult due to overlapped spectra. The hexadecane concentration was carried out by using the maximum peak in the 2984–2833 cm^{-1} range. The variation of hexadecane concentration with the microplasma process time is shown in **Figure 6**. The values are lower than in the case of natural decay. From an initial concentration of 3000 ppm, the hexadecane decreased to 150 ppm after the microplasma process for 120 minutes. By comparison the natural decay was to 950 ppm after 120 minutes; thus microplasma decomposed the hexadecane. The removal energy efficiency of hexadecane was calculated at 19.8 μg/W h. Eqs. (1)–(3) explain the hexadecane decomposition process. Microplasma is a nonequilibrium plasma under atmospheric pressure; thus active species and high-energy electrons generated by microplasma decompose the hexadecane [38]. At the interaction with an electron, the hexadecane breaks in two neutral radical molecules:

$$C_{16}H_{34} + e^- \leftrightarrow 2C_8H^*_{17} + e^- \quad (1)$$

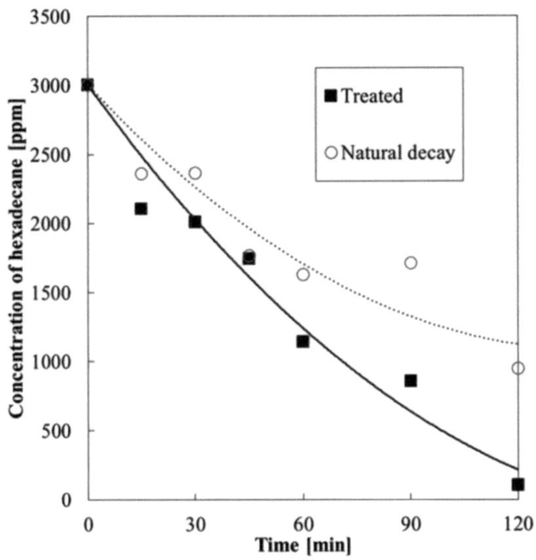

Figure 6.
Concentration of hexadecane with processing time.

Anionic or cationic carbon species could be also obtained after electron impact:

$$C_{16}H_{34} + e^- \leftrightarrow C_8H^-_{17} + C_8H^+_{17} + e^- \quad (2)$$

The loss of a proton generates alkene:

$$C_8H^+_{17} \leftrightarrow C_8H_{16} + H^+ \quad (3)$$

Due to the active species generated by microplasma, the carbon single bonds were cut, and substances that have less carbon atoms were formed as intermediate products. To conclude, after the microplasma process of hexadecane removal, the following results were obtained:

1. The hexadecane concentration was decreased from the initial value of 3000 ppm to 150 ppm after the 120 minutes of microplasma treatment. In comparison, the natural decay was to 950 ppm; thus microplasma decomposed the hexadecane.

2. The removal energy efficiency of hexadecane was calculated at 19.8 μg/W h.

2.2 Skin treatment by plasma for drug delivery

A new method of drug delivery through the skin is plasma by modifying the skin permeability [12–15, 39–41]. Researchers were investigating the phenomena of interaction between the plasma and the skin. The penetration of the reactive oxygen species (ROS) using plasma jet with discharge gas helium was investigated by Szili et al. [42]. Another study where bovine keratin taken as a model protein was subject to dielectric barrier discharge plasma treatment and showed the generation of cysteic acid by oxidation of disulfide bonds accompanied by a decrease in pH [43]. The human forearm was treated by dielectric barrier discharge plasma and an increased number of C—O and C=O bonds, and a decreased number of C—C bonds in the lipids were measured after the treatment as reported by Marschewski et al. [44]. Various processes were used for the enhancement of skin permeability such as chemical solutions, ablation, iontophoresis, or electroporation [45].

The influence of argon plasma on the thickness on the stratum corneum layer and extraction of lipids from stratum corneum was investigated. Attenuated total reflectance-Fourier transform infrared (ATR-FTIR) spectroscopy and section measurement of skin were used for measuring the modifications caused by plasma in the stratum corneum layer. Only lipophilic molecules and molecules with weight less than 500 g/mol can penetrate through the skin because of the stratum corneum, a lipid-rich matrix with embedded corneocyte cells. The majority of active agents for therapeutic applications are represented by relatively large drugs; thus the low permeability is problematic. Skin permeability was studied using attenuated total reflectance-Fourier transform infrared spectrometry. Information about the chemical structure of lipids and their amount in the stratum corneum could be obtained by studying the spectra of the skin. Also, the microscopic observation could give information about the morphologic changes across a section of the skin. The etching of the skin, lipid extraction, and appearance of the skin after atmospheric pressure argon plasma treatment were investigated.

2.2.1 Experimental setup

The schematic image of the experimental setup using the plasma jet is shown in **Figure 7(a)**. Nonthermal plasma was generated at atmospheric pressure by a plasma jet type structure consisting of a Pyrex tube (outer diameter 6 mm, length 100 mm) and a central tungsten (0.8 mm in diameter) high-voltage (HV) electrode covered by a glass layer except for a 10-mm long region close to outlet of Pyrex tube. The grounded electrode was an aluminum ring (8 mm diameter) located on the outer surface of the Pyrex tube at 12 mm from the end of the plasma jet. The electrodes were energized using a Neon transformer (ALPHA Neon M-5) with an AC voltage with frequency of 16 kHz. The voltage and current were measured using a Tektronix P60015A high-voltage probe and a Tektronix P6021 current probe. A grounded

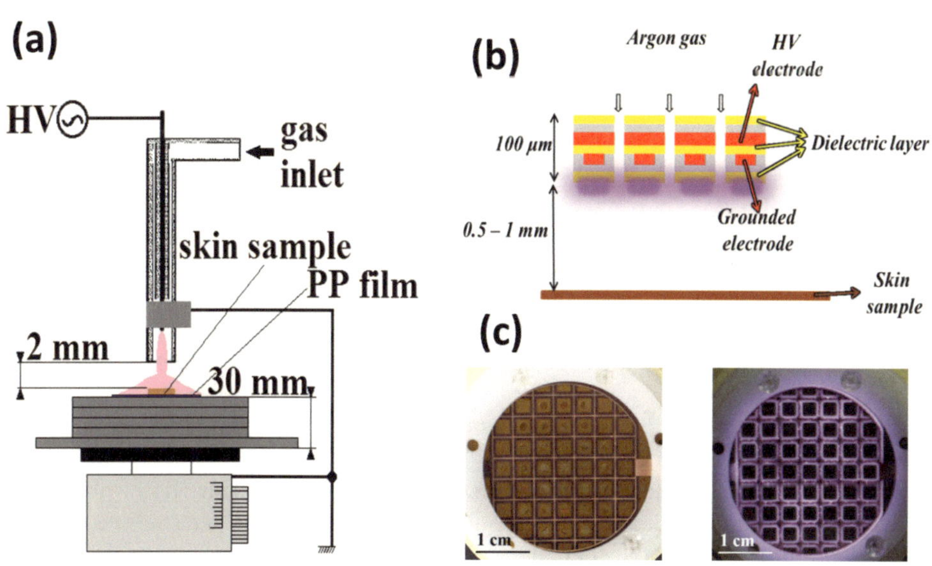

Figure 7.
Experimental setup for plasma treatment of skin: (a) plasma jet; (b) microplasma and (c) photo of microplasma electrode without and with discharge.

micrometric sample holder was used to set the distance between the skin sample and the outlet of the plasma jet at 2 mm, and a 30-mm thick PVC insulator was used to insulate the sample from the holder.

The plasma treatment was carried out for 1 minute, and the sample was in contact with plasma during this time. Argon gas (purity 99.99%) was flowed through a Yamato flow meter in to the plasma jet. The upper layer of stratum corneum of pig skin was observed by attenuated total reflectance-Fourier transform infrared (ATR-FTIR) spectrometer (Jasco FT/IR 6300 with ATR PRO610P-S) with a diamond prism, and the spectra were recorded with a resolution of 8 cm^{-1} and by accumulating 150 scans.

Yucatan micropig skin from Charles River Laboratories Japan, Inc. (Yokohama, Japan) was used for experiments of the plasma treatment. The skin was stored at −20°C in a freezer before the experiments. The pig skin preparation procedure was first, the fat layer of the skin was removed by using a knife, then it was cut and soaked at 4°C in phosphate buffered saline (PBS) for 3 hours, and then after a bath in 60°C PBS for 1 minute, the epidermal layer of a thickness of 200 μm was peeled using tweezers. Finally the skin sample was cut in to 3 × 3 mm parts and attached to a polypropylene film by using double-sided tape.

A dielectric barrier discharge microplasma was generated by a thin-film electrode with a thickness of 105 μm is shown in **Figure 7(b)** and **(c)**. Both high-voltage and grounded electrodes were encapsulated in the dielectric. The thickness of electrodes was 18 μm separated by a 14 μm thick dielectric. The outer side of each electrode was covered by a dielectric with a thickness of 27.5 μm. Plasma was generated at the surface closer to the grounded electrode. Atmospheric argon plasma was generated at a voltage of 950 V and a frequency of 25 kHz using a Neon transformer (ALPHA Neon M-5, LECIP). Gas flow of argon was set to 5 l/minute using a flow meter (Yamato). The treatment time of the skin was set to 5 minutes after the gas flow was started. The thickness of the plasma was negligible in comparison with the distance between the electrode and the skin surface, which was set to 0.5–1 mm. Thus, the plasma was not in contact with the skin.

2.2.2 Skin treatment by plasma for transdermal drug delivery

A microscope was used to visualize a microscale change in the stratum corneum; thus the paraffin-embedded skin samples were sectioned at a thickness of several micrometre using a microtome and were placed on a glass slide.

Figure 8a shows the untreated sample, and **Figure 8b–d** shows the images of sample treated for 10 s, 30 s, and 60 s at a voltage of 4 kV amplitude. A disruption of the stratum corneum layer on the treated samples is observed. The upper layer of stratum corneum was detached and the thickness decreased.

The ATR-FTIR spectra of the pig skin stratum corneum samples before and after the plasma jet treatment were compared.

The asymmetric stretching band of CH_2 (2917–2923 cm^{-1}) was examined due to a highest absorbance of the band. The electrodes were energized at 4 kV amplitude, and the skin samples were measured at various treatment times from 10 s to 15 minutes.

The extraction of lipids with treatment time and saturation of their extraction after 120 s are shown in **Figure 9**.

The thickness of the stratum corneum decreased to 11 μm after 60 s of plasma treatment. The skin barrier effect depends on the thickness of the stratum corneum and the amount of lipids and their structure; thus the barrier function of the skin was reduced. The etching effect could be attributed to the ions or radicals created in the plasma.

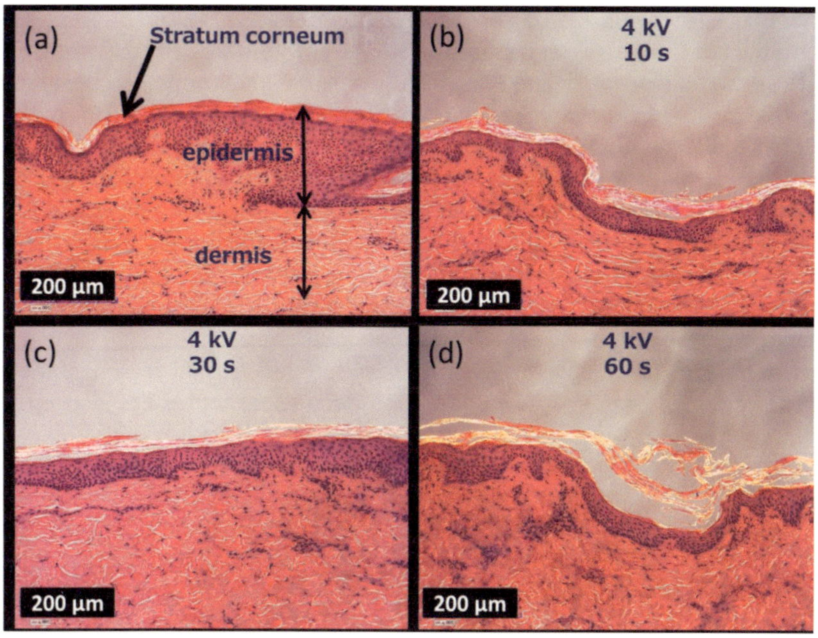

Figure 8.
Section of skin samples. (a) Untreated sample, (b) dielectric barrier discharge argon plasma treatment for 10 s at 4 kV, (c) dielectric barrier discharge argon plasma treatment for 30 s at 4 kV, and (d) dielectric barrier discharge argon plasma treatment for 60 s at 4 kV.

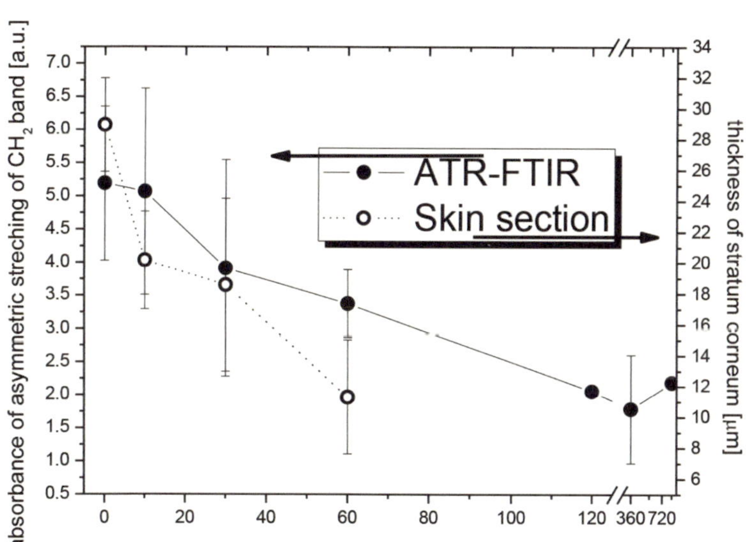

Figure 9.
Evolution of absorbance of the asymmetric stretching band of CH_2 after various times of treatment (black circles). Evolution of thickness of stratum corneum (empty circles).

Because of skin regeneration, the upper cells of the skin are attached weaker than the deeper layers; thus the etching effect can be nonlinear. The extraction of lipids and its saturation are different for the layers which are not in contact with plasma. A relatively long-term effect of plasma was measured, and the description of a fast effect of plasma treatment that can increase permeability of the skin was not carried out.

After the experiments of atmospheric pressure plasma jet treatment of pig skin, the disruption of stratum corneum was measured. It was found that after 30 s of treatment, the skin was disrupted, and ATR-FTIR measurements confirmed the extraction and also the saturation of etching after 120 s. The thickness of the stratum corneum was decreased to one third (11 μm) of the original thickness after 60 s of treatment.

The treatment of the skin by dielectric barrier discharge microplasma was carried out. Full-thickness (dermal + epidermal layer) skin was cut to ultrathin sheets by ultramicrotome to observe the lamellar structure of the lipid matrix. This matrix can be observed as bright lines with dark space in between. Bright lines represent lipid tails and dark spaces represent hydrophilic heads [46]. Unaltered lipid lamellae are shown in **Figure 10**.

Lipid lamellae of microplasma-treated skin shows disturbed areas (nonhomogeneous) of lipid membrane with lacunar space along corneocytes and also extracted

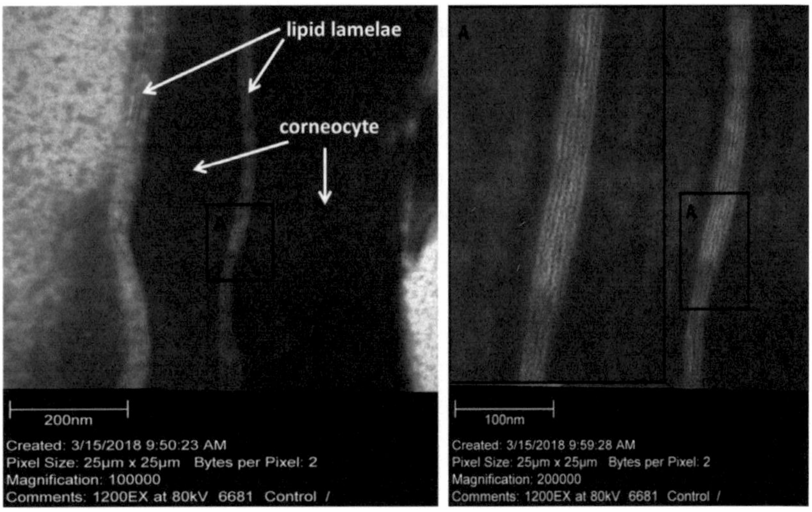

Figure 10.
Lipid lamellae of untreated skin sample with detail in A.

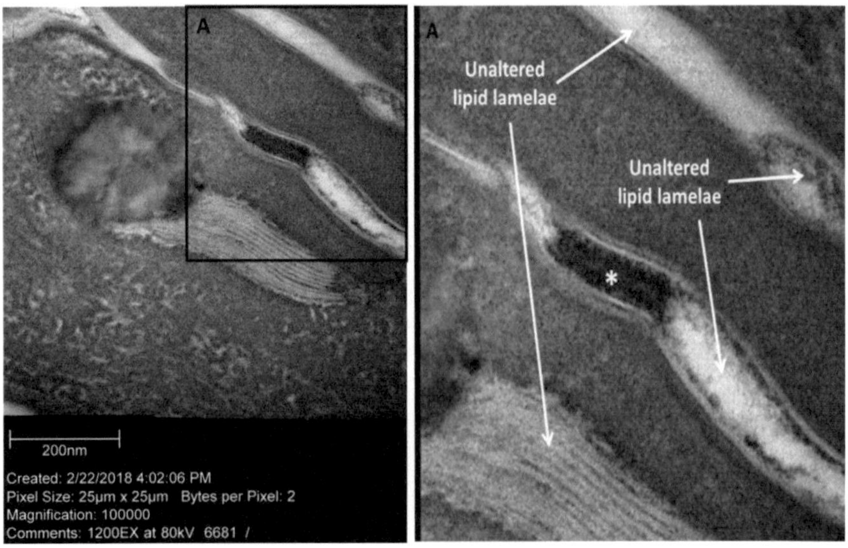

Figure 11.
Lipid lamellae of microplasma-treated skin sample with detail in A. Asterix—space without lipid lamellae.

lipids in "asterix" space in **Figure 11**. The skin was treated for 5 minutes by argon microplasma.

The results obtained using plasma jet and microplasma showed the modification of the skin in order to carry out the transdermal drug delivery.

2.3 Particle removal

Plasma technology was applied for the removal or collection of fine particles. Microplasma electrode was used to remove the fine particles from the electrode surface due to the action of electrostatic forces. Microplasma [47] was already proved to be an economical and ecological solution in the case of biomedical applications [1, 48], surface treatment [49], material synthesis [50], and indoor air treatment [33]. The removal or transfer of fine particles is important for global environmental problems such as PM 2.5, semiconductor industry dust contamination [16], or for the exploration activities on the moon [17, 18]. Research results regarding the dust removal process using dielectrophoretic and electrostatic forces were published by Calle et al. [18] and Kawamoto et al. [19]. Because the charged or uncharged lunar dust has a negative influence on the operation of mechanical system, lunar dust simulant was used in the series of experiments. Also, other researchers investigated removal of the fine particles from the surface of solar cells [20, 21], the transport of toner particles by traveling wave [51, 52] and the decontamination of the silicon wafer [53, 54]. Due to its advantages such as low discharge voltage of about 1 kV and scalability of electrode size, microplasma was used for particle removal [55]. An AC voltage with amplitude 1 kV was applied to microplasma electrodes in order to remove SiO_2 particles from their surface. The dependency of the particle removal with the frequency of applied voltage was investigated.

2.3.1 Microplasma electrode for particle removal

For particle removal experiments, the microplasma electrode shown in **Figure 12** was used [55].

The electrode structure consists of strip-type electrodes (top-side electrode), a plate-type electrode (bottom-side electrode), a dielectric layer, and an isolating layer.

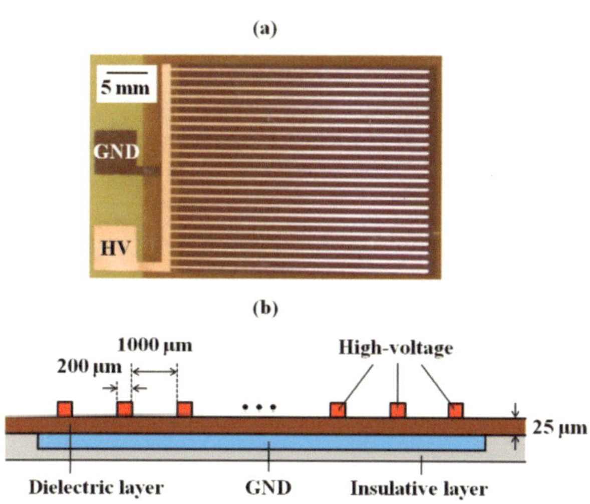

Figure 12.
Schematic images of the microplasma electrode: (a) top view and (b) cross section view.

For the stripe-type electrodes, copper was used and as a dielectric layer polyimide. The microplasma electrode has a capacitive load of 200 pF. The top-side electrode consisted of 23 stripes with 1 mm in between and was energized at high-voltage (HV), and the plate type bottom-side electrode was grounded. The top strip-type electrodes' width and length were 200 μm and 30 mm, respectively. The plate-type electrode area was 23 × 30 mm^2. The dielectric in between the HV electrode and grounded electrode had 25 μm thickness; thus plasma could be generated at a discharge voltage of about 1 kV. The experimental setup for the fine particle movement experiments is shown in **Figure 13**. The HV electrode was energized by an AC voltage with 1 kV amplitude, and the frequency was varied from 10 Hz to 10 kHz. The voltage was obtained using a function generator (Tektronix AFG3102) and a HV power amplifier (TREK 5/80).

A high-voltage probe (Iwatsu HV-P30) and an oscilloscope (Tektronix TDS 2014) were used for the measurement of applied voltage. For the measurements of the discharge current, a resistor was connected between the electrode and ground. The voltage drop on the resistor was measured using a voltage probe (Tektronix P2220) and an oscilloscope (Tektronix TDS 2014), and thus the current was obtained by Ohm's law. The particles used in the series of experiments were SiO_2 particles (diameter, 50 μm; density, 2.2 g = cm^3; and shape, spherical).

For each experiment, 60 mg of SiO_2 particles were placed on the electrode surface before applying the AC voltage (applied to the electrodes for 3 minutes). The waveforms of applied voltage and discharge current are shown in **Figure 14** for the frequency of the applied voltage 50 Hz, 500 Hz, and 5 kHz.

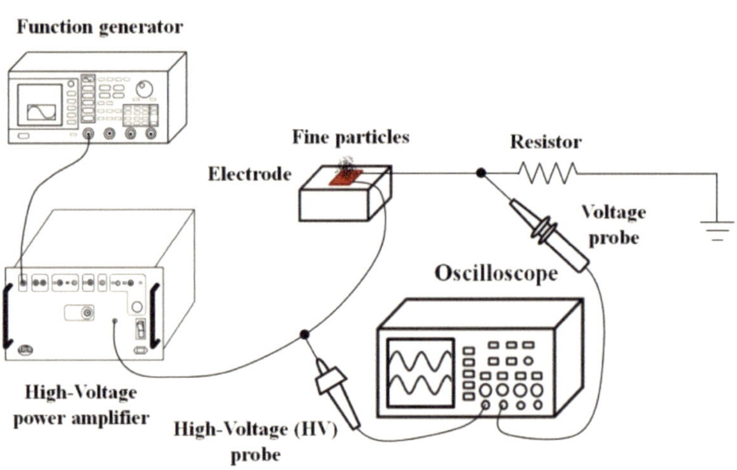

Figure 13.
Experimental setup for observing fine particle movement.

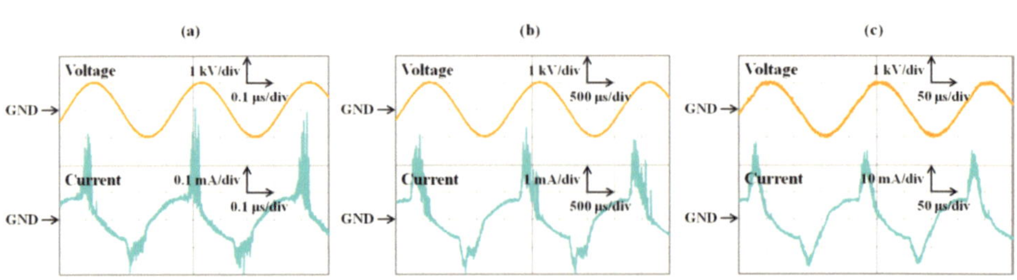

Figure 14.
Waveforms of the applied voltage and the discharge current at (a) 50 Hz, (b) 500 Hz, and (c) 5 kHz.

The discharge current had two components: the capacitive current (due to the capacitance of the electrode, its phase led that of the applied voltage by about 90°) and the current spike. The streamer propagation and glow-like discharge caused the current spikes during positive- and negative-going cycles. The intensity of the current spikes during the positive-going cycle was larger than that during negative-going cycle in all cases due to the fact that the bottom-side electrode was buried.

It was also confirmed that the number of current spikes increased with the frequency of the applied voltage. The power dissipated by the microplasma electrode was measured using the charge-voltage curve (Lissajous figure). The Lissajous figure corresponding to an AC voltage with amplitude 1 kV and 1 kHz is shown in **Figure 15**. The measured power consumption was estimated at 0.27 W. At lower frequencies in the 10–100 Hz range, the Lissajous figure could not be obtained, thus power consumption was estimated by multiplying the applied voltage and discharge current.

The power consumption versus the frequency of the applied voltage is shown in **Figure 16**. At 1 kHz the power consumption was calculated to be 0.28 W which was almost the same value calculated using the Lissajous figure.

With the increase of frequency, the power consumption increased. This can be explained by the dielectric barrier discharge phenomena: discharges occurred at the time the polarity of the voltage was inverted, thus discharges per unit time increased.

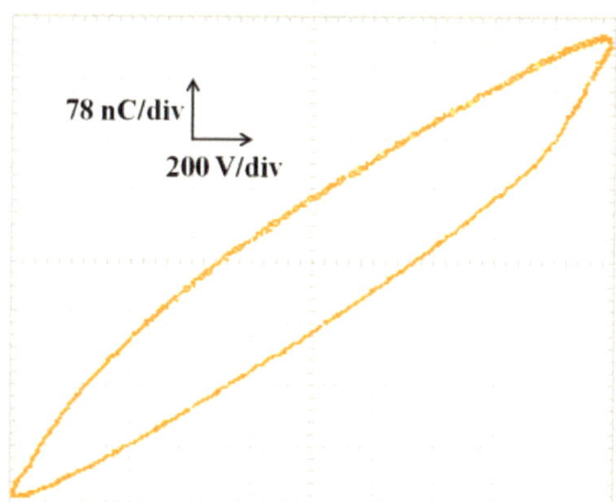

Figure 15.
Lissajous figure when AC voltage with a voltage of 1 kV and a frequency of 1 kHz was applied to the electrode.

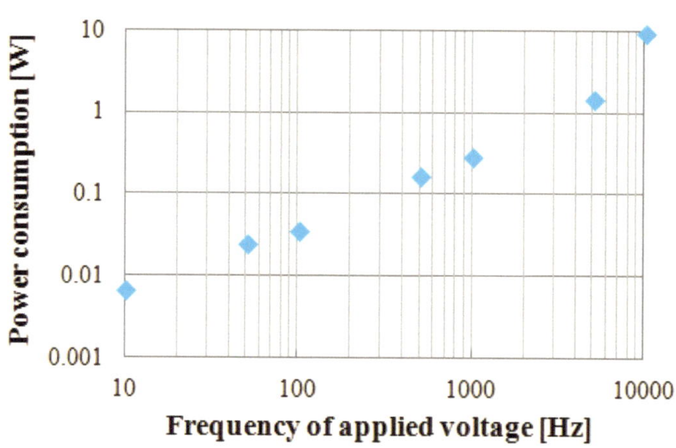

Figure 16.
Power consumption as a function of the frequency of the applied voltage.

2.3.2 Particle removal by microplasma

The SiO_2 particle removal process was investigated when the electrode was energized at an AC voltage with a frequency of 10 Hz, 100 Hz, 1 kHz, and 10 kHz as shown in **Figure 17**. After 5 s, the number of removed particles increased as the frequency increased because more ions were generated at high frequency. With the increase of process time the amount of removed particles decreased as the frequency of the applied voltage increased. Some amount of particles still remained on the electrode surface at high frequency such as 1 and 10 kHz because at high frequency, movement of fine particles could not follow the high-speed inversion of

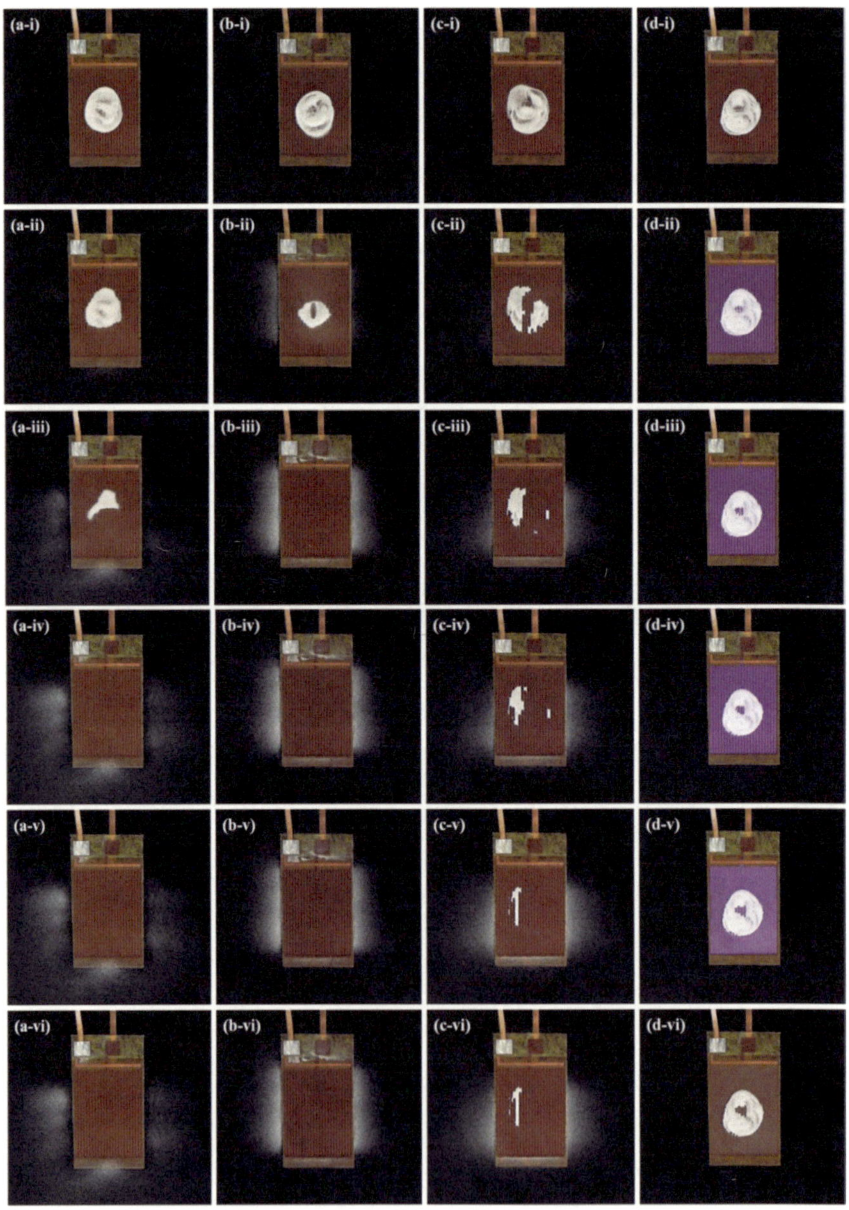

Figure 17.
SiO_2 particle movement when the frequencies were set to (a) 10 Hz, (b) 100 Hz, (c) 1 kHz, and (d) 10 kHz: (i) before applying the voltage, after applying voltage for (ii) 1 s, (iii) 5 s, (iv) 10 s, (v) 1 minute, and (vi) 3 minutes.

the polarity. The same phenomena was reported by Kawamoto et al. [19, 20] which carried out fine particle removal and transport and reported that the cleaning efficiency decreased at more than 10 Hz. The threshold value reported in [19, 20] was different from these results obtained at 500 Hz due to the different conditions such as electrode configuration, applied voltage, particle diameter, and electrical properties of particles. The removal rate was defined as

Removal rate [%] = (The mass of the particles removed to outside part)/(The mass of the particles placed on the electrode (60 mg)) × 100 (4)

The removal rate at various frequencies is shown in **Figure 18**. Each value represents the mean value obtained after three experiments at each frequency. The removal rate for frequencies between 10 and 500 Hz was about 95%, and for frequencies above 500 Hz, the removal rate decreased with the increase of frequency; thus 72% was measured at 1 kHz and 17% at 10 kHz, respectively. Considering also the power consumption, the removal efficiency was high at low frequency.

The process of particle transfer could be described as:

i. Microplasma was generated at the electrode surface by applying the AC high voltage to the electrode.

ii. The electrical charging of SiO_2 particles was due to the contact with the electrode and microplasma.

iii. The Coulomb and dielectrophoretic forces acted on the charged SiO_2 particles.

iv. With every high-voltage cycle, the processes (ii) and (iii) were repeated, thus SiO_2 particles were transferred to the outside part of the electrode.

After the series of experiments, the following conclusions were obtained:

1. The power consumption of the microplasma electrodes increased with the increase of the frequency of the applied voltage.

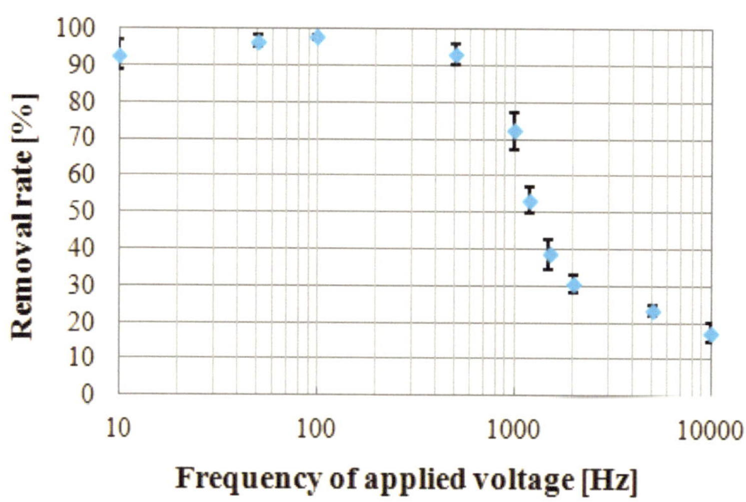

Figure 18.
Removal rate as a function of the frequency of the applied voltage.

2. The removal rate of SiO_2 particles decreased with the increase of frequency. For frequencies between 10 and 500 Hz, the removal rate was about 95%, but the removal rate decreased when the frequency was 500 Hz or above. At 10 kHz 83% particles remained on the electrode surface due to the fact that the SiO_2 particles movement could not follow the polarity changes.

2.4 Microplasma actuator for flow control

Plasma actuators could be a safer and more efficient alternative to conventional mechanical actuators used on aircraft due to their lack of moving parts, simple construction, and fast response due to the electric field [56]. A general classification could split plasma actuators in dielectric barrier discharge (DBD) type and corona discharge type. The induced flow values could reach 7 m/s in the case of single DBD plasma actuator and 11 m/s when a multiple DBD plasma actuator was used [24]. Corona discharge plasma actuators could be used for high-speed flow control where an induced flow speed of more than 7 m/s is required [57–61]. High voltages of about 7 kV peak-to-peak are required for energizing the millimeter-sized discharge gap of DBD plasma actuators applied to turbulent boundary layer control for skin friction drag reduction [62–65]. Research has been done on dielectric barrier discharge (DBD) plasma actuators. This type of plasma actuator is conventionally energized at voltages of tens of kilovolts [22–24]. For flow control applications, a microplasma actuator which can be energized at only 1 kV was developed; thus only a smaller-sized power supply is necessary to energize the actuator, and there are no difficulties of electrically insulating the high voltage.

The induced flow by microplasma actuator occurs due to electrohydrodynamic (EHD) phenomenon through which by collision the momentum is transferred from ions accelerated by the electric field to neutral molecules [22–24].

Microplasma actuator has small dimensions; thus the EHD flow measurements are difficult to carry out. Besides the experimental investigation, numerical simulation software was developed to study the microplasma actuator phenomena especially near the electrode surface where, due to the light emission, the measurements were difficult [25–28].

For the numerical simulation of the flow, the Suzen-Huang model coupled with Navier-Stokes equations was used [66–68]. Other researchers obtained numerical simulation results close to the physical phenomena of the plasma actuator using the Suzen-Huang model [69–71].

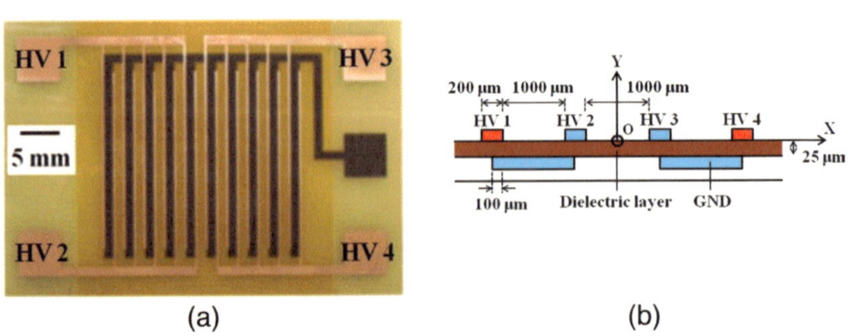

Figure 19.
Multi-electrode microplasma actuator. (a) Top view showing strip-like electrodes: HV as energized electrodes and GND as ground; (b) Schematic cross-section view.

2.4.1 Microplasma actuator

The schematic image of microplasma actuator is shown in **Figure 19**. It has a multielectrode structure which allows various parts of the actuator to be energized at different potentials or waveforms, thus obtaining different flow patterns.

The energized top-side electrode has 20 strip-like electrodes with a width of 200 μm and thickness of 16 μm made from copper and placed above a plate type grounded electrode also made from copper. A dielectric layer of resin film with a thickness of 25 μm is placed in between the strip-like electrodes and plate type electrode.

Microplasma actuator is composed of four active parts as shown in **Figure 19**, which can be driven independently. In the series of experiments, an AC voltage with amplitude 1.4 kV and frequency 20 kHz was applied to the parts HV1 and HV4. The applied voltage was pulse-modulated with 50% duty ratio and a burst frequency of 4 kHz. Parts HV1 and HV4 were energized by the AC voltage for 50 ms.

For the flow measurements using the particle tracking velocimetry method (PTV), incense smoke composed from 0.3 μm tracer particles and a laser (Nd YVO_4 532 nm) were used. The induced flow by the microplasma actuator was measured with a high-speed camera of 1280 × 128 pixel resolution. The recording frequency of the high-speed camera was at 4000 Hz. The PTV results were obtained using two consecutive images.

2.4.2 Flow control by microplasma actuator

The active electrode parts HV1 and HV4 were energized by a pulse-modulated AC voltage with amplitude 1.4 kV, frequency 20 kHz, and duty ratio 50%.

As shown in **Figure 20**, at time t = 1 ms, the vortexes appeared near the strip-like energized electrodes having a counterclockwise direction for the HV1 part and clockwise direction for the HV4 part.

Gradually in time the vortexes above part HV1 and part HV4 unite and form a flow directed toward the right for the HV1 part and toward left above the HV4 part. Furthermore, these two flows collide in the center part of the actuator, and upward flow is formed as shown in **Figure 20** at t = 10 ms. The PTV results from **Figure 21**

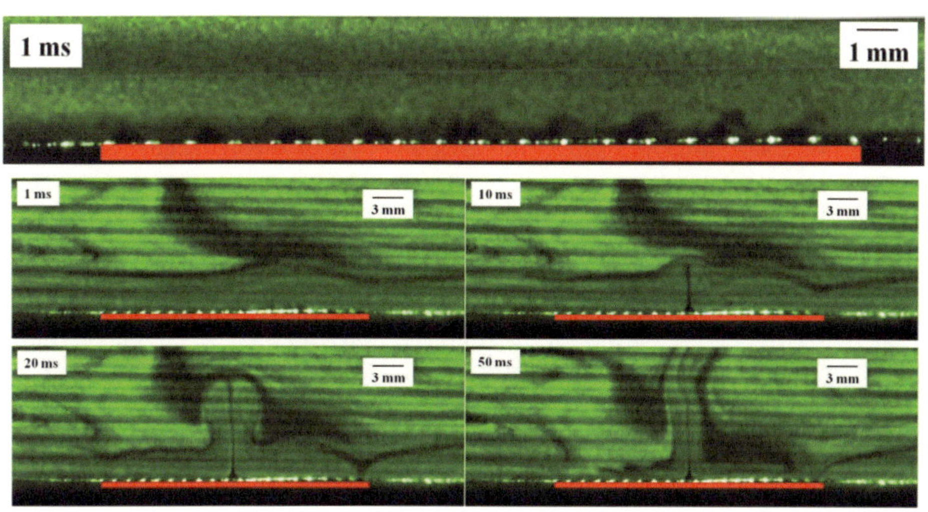

Figure 20.
Images of generated flow by microplasma actuator. Up to 10 ms vortexes appeared above the active electrodes. After 10 ms upward flow is obtained.

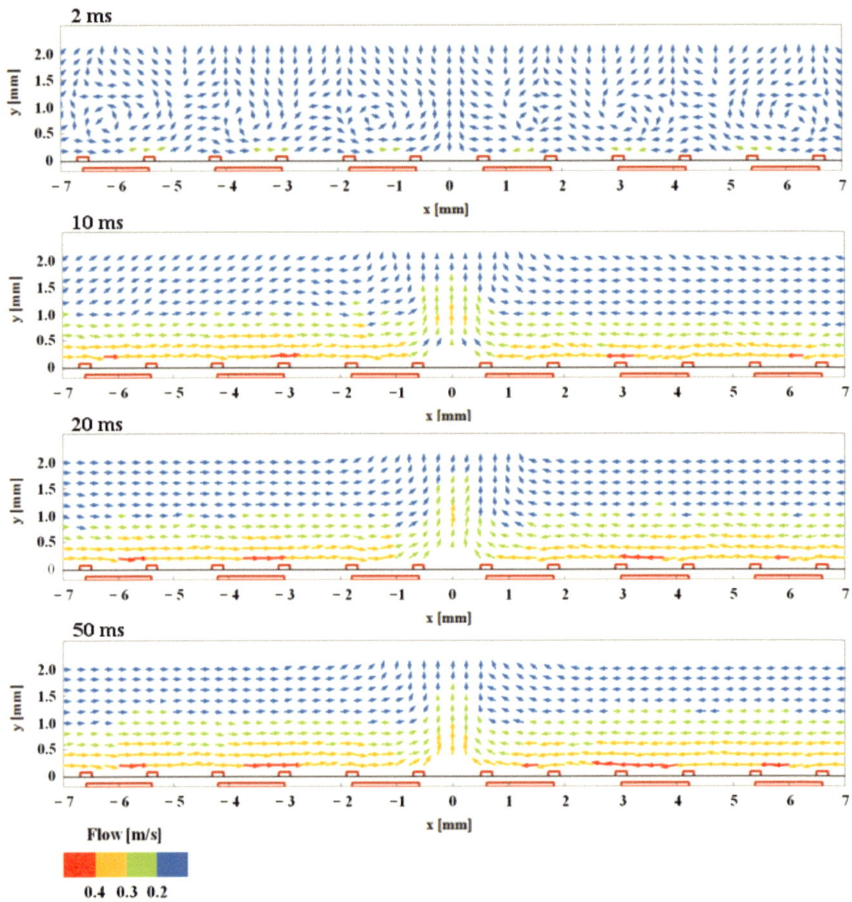

Figure 21.
PTV results of generated flow by microplasma actuator. The maximum flow velocity was measured above the active electrodes.

show the maximum values of flow velocity near the electrodes which at t = 2 ms was about 0.3 m/s and at t = 10 ms was about 0.5 m/s.

In the center part of the actuator the maximum value was about 0.4 m/s.

Near the active electrodes due to the microplasma light emission, the measurements were difficult to carry out; thus numerical simulations could give additional information about the flow. Numerical simulation code based on the Suzen-Huang model [66–68] was developed. Microplasma is an atmospheric pressure weakly ionized nonthermal plasma; thus the model considers the potential V to be composed from two potentials: the external electric field potential Ø and the net charge density potential φ. Furthermore the body force was calculated based on these potentials.

Figure 22 shows the geometry used for numerical simulation. For the dielectric the permittivity was $\varepsilon r_d = 4$ and for the air the permittivity was $\varepsilon r_{air} = 1$. Furthermore the harmonic mean of the permittivity was calculated for the conservation of the electric field [67]. The grounded electrodes were considered as source charge, and the value of the source charge was $\rho c = 0.00752\ C/m^3$. The value of Debye length was $\lambda D = 0.00017$ m for the air and $\lambda D = \infty$ for the dielectric [67]. In order to obtain the flow, the calculated body force Fx and Fy on the x- and y-axes ,respectively, were introduced in the Navier-Stokes equations.

The flow velocity has the components u and v on x-axes and y-axes, respectively. The dynamic viscosity μ was calculated with the value of air density $\rho = 1.177\ kg/m^3$

and the value of kinematic viscosity $\upsilon = 1.57 \times 10^{-5}\ m^2/s$; thus dynamic viscosity was $\mu = 1.8 \times 10^{-5}$ kg/m s.

There was no wall on the left, right, and upper part.

The numerical calculations were carried out for six active electrodes (three for HV1 and three for HV4) with grid dimensions 14 × 14 mm and 561 × 561 grid nodes as shown in **Figure 22**.

The active electrodes were energized by an AC waveform with amplitude 1 kV, 20 kHz, and duty ratio 50%. The body force magnitude and direction according to the "push-push" theory are similar for the positive and negative half-cycles [68]. Because the effective value of AC voltage with 1.4 kV amplitude is 1 kV, we have considered for the numerical simulation a positive pulse signal with peak value of 1 kV with duty ratio 50% as shown in **Figure 23**.

For the equations of potentials and body force which were solved before using the Navier-Stokes equations, the finite difference method was used for the discretization. The calculated body force had the highest intensity near the active electrodes.

Vortexes appeared in the initial stages of discharge near the energized electrodes as shown in **Figure 24**. The vortexes were rotating counterclockwise for the Ch1 electrodes and clockwise for the Ch4 electrodes up to 10 ms. With the lapse of time, after 10 ms the vortexes start to unite on the left side and right side of the microplasma actuator; thus over the Ch1 electrodes, the resultant flow was rightward flow, and leftward flow was obtained on the right side of the actuator over the Ch4 electrodes. The obtained flows collide in the center part of the actuator and an upward flow resulted in that region. The maximum value of the flow was about 0.5 m/s and was calculated near the active electrodes at 1 ms. At 10 ms the maximum value increased to about 0.7 m/s. In the center region the upward flow maximum value was about 0.45 m/s and was obtained after 20 ms. The calculation was carried out up to 50 ms when it can be observed that the upward flow extends up to the upper part of the calculation geometry boundary. At up to 0.2 mm height from the active electrodes, the microplasma light emission influenced the measurement; thus the experimental data in that region was not accurate. The numerical results gave us additional information about the flow near the active electrodes region.

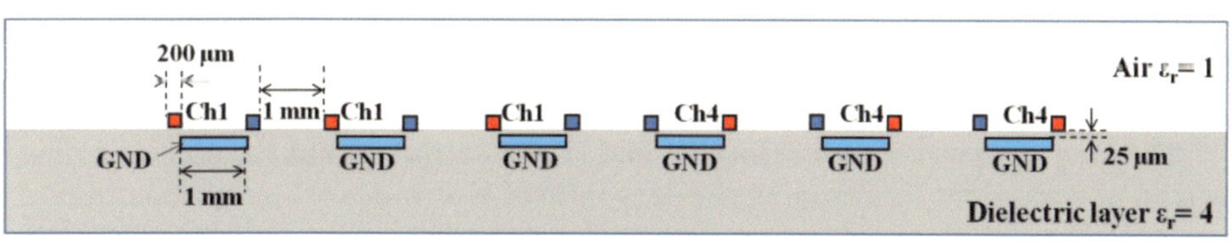

Figure 22.
Computational geometry. Ch1 and Ch4 electrodes were energized.

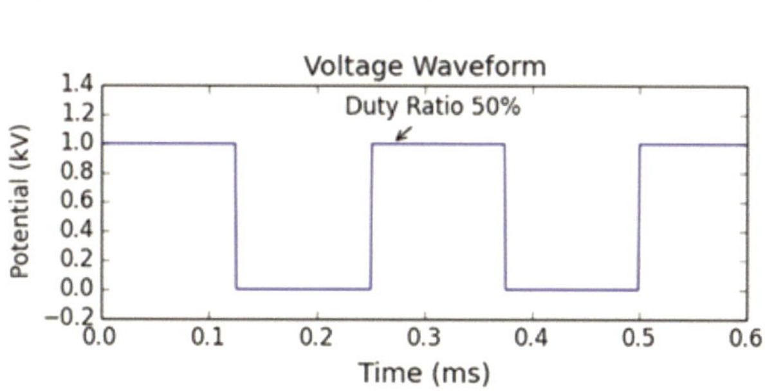

Figure 23.
Voltage waveform.

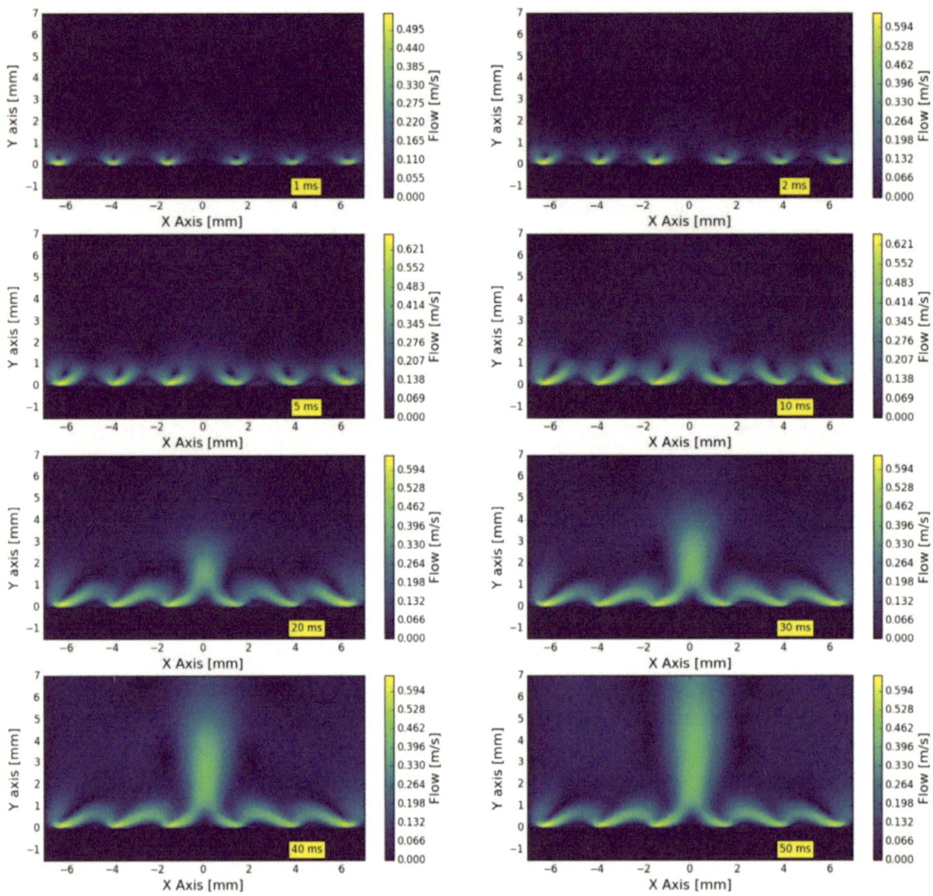

Figure 24.
Induced flow by the microplasma actuator.

The comparison between the experimental results and numerical simulation results shows a good fit considering the flow characteristics and values. The numerical simulation data was more accurate than the experimental results near the active electrodes region.

The duty ratio was 50% for the HV1 and HV4 electrodes, thus the obtained flow was an upward flow. By changing the duty ratio of electrodes or energizing other parts (HV2 or HV3), the flow direction can be changed.

3. Conclusions

Dielectric barrier discharge microplasma has proven to be a technology with various applications that can replace classical technologies for indoor air purification, drug delivery, particles removal and induced flow. Microplasma can be generated at voltages of about 1 kV making this technology safer, cheaper and easier to insulate and integrate in other systems compared with classical non-thermal plasma technologies which requires tens of kilovolts. The low discharge voltage requires a smaller-sized power supplies which also contributes to the miniaturization of the system. The electrodes can be scaled up to large areas depending on the applications requirements without increasing the discharge voltage and in the case of microplasma actuator the flexible-type electrode can be bent in various shapes to the surface.

The microplasma process of hexadecane removal decreased the concentration of hexadecane from the initial value of 3000 ppm to 150 ppm after the 120 minutes of microplasma treatment. In comparison, the natural decay was decreasing to 950 ppm.

The plasma treatment of pig skin decreased to one-third (11 μm) that of the original thickness after 60 s of treatment. After 30 s of treatment, the skin was disrupted, and the saturation of etching effect was observed after 120 s.

The removal of SiO_2 particles from the surface of microplasma electrode decreased with the increase of frequency. For frequencies between 10 and 500 Hz, the removal rate was about 95%, but the removal rate decreased when the frequency was 500 Hz or above.

The microplasma actuator induced an upward flow, while the electrode was energized with a duty ratio of 50% in both left part and right part of the actuator. The simulation results were confirmed by experimental results.

Acknowledgements

The authors would like to thank Professor Hitoki Yoneda from the University of Electro-Communications, Tokyo, Professor Damon Chandler, Mr. Yoshinori Mizuno, Mr. Akihiko Ito, and Mr. Daisuke Nonaka from Shizuoka University, Hamamatsu, for the fruitful discussions.

Author details

Kazuo Shimizu*, Jaroslav Kristof and Marius Gabriel Blajan
Shizuoka University, Hamamatsu, Japan

*Address all correspondence to: shimizu@cjr.shizuoka.ac.jp

References

[1] Iza F, Kim GJ, Lee SM, Lee JK, Walsh JL, Zhang YT, et al. Microplasmas: Sources, particle kinetics, and biomedical applications. Plasma Processes and Polymers. 2008;**5**(4):322-344

[2] Foest R, Schmidt M, Becker K. Microplasmas: An emerging field of low-temperature plasma science and technology. International Journal of Mass Spectrometry. 2006;**248**(3):87-102

[3] Shimizu K, Ishii T, Blajan M. Emission spectroscopy of pulsed power microplasma for atmospheric pollution control. IEEE Transactions on Industry Applications. 2010;**46**(3):1125-1131

[4] Blajan M, Umeda A, Muramatsu S, Shimizu K. Emission spectroscopy of pulsed powered microplasma for surface treatment of PEN film. IEEE Transactions on Industry Applications. 2011;**47**(3):1100-1107

[5] Orosa J. Indoor and Outdoor Air Pollution. Rijeka, Croatia: InTech; 2011

[6] Arashidani K, Akiyama Y, Kunugita N. Study of indoor air pollution. Indoor Environment. 2009;**12**(2):71-86

[7] Weschler CJ. Changes in indoor pollutants since the 1950s. Atmospheric Environment. 2009;**43**:153-169

[8] Eliasson B, Hirth M, Kogelschatz U. Ozone synthesis from oxygen in dielectric barrier discharges. Journal of Physics D: Applied Physics. 1987;**20**(11):1421-1437

[9] Lerouge S, Fozza AC, Wertheimer MR, Marchand R, Yahia LH. Sterilization by low-pressure plasma: The role of vacuum-ultraviolet radiation. Plasmas and Polymers. 2000;**5**(1):31-46

[10] Moisan M, Barbeau J, Moreau S, Pelletier J, Tabrizian M, Yahia L'H. Low-temperature sterilization using gas plasmas: A review of the experiments and an analysis of the inactivation mechanisms. International Journal of Pharmaceutics. 2001;**226**:1-21

[11] Shimizu K. Indoor air control by microplasma. In: Farhad Nejadkoorki, editor. Advanced Air Pollution. IntechOpen; 2011. pp. 509-535. DOI: 10.5772/16764. Available from: https://www.intechopen.com/books/advanced-air-pollution/indoor-air-control-by-microplasma

[12] Lademann J, Patzelt A, Richter H, Lademann O, Baier G, Breucker L, et al. Nanocapsules for drug delivery through the skin barrier by tissue-tolerable plasma. Laser Physics Letters. 2013;**10**:083001

[13] Shimizu K, Hayashida K, Blajan M. Novel method to improve transdermal drug delivery by atmospheric microplasma irradiation. Biointerphases. 2015;**10**:029517

[14] Kalghatgi S, Tsai C, Gray R, Pappas D. Transdermal drug delivery using cold plasmas. In: 22nd International Symposium on Plasma Chemistry. July 5-10, 2015

[15] Kristof J, An NT, Miyamoto H, Blajan MG, Shimizu K. Skin permeability and transdermal drug delivery by plasma irradiation. In: Proc. 18th Takayanagi Kenjiro Memorial Symposium. Hamamatsu, Japan: Shizuoka Univ.; Nov. 15-16, 2016

[16] Hattori T. Contamination control: Problems and prospects. Solid State Technology. 1990;**33**(7):S1-S8

[17] Christoffersen R, Lindsay JF, Noble SK, Meador MA, Kosmo JJ, Lawrence JA, et al. NASA=TP-2009-214786; 2009

[18] Calle CI, McFall JL, Buhler CR, Snyder SJ, Arens EE, Chen A, et al. Proc. ESA Annu. Meet. Electrostatics; 2008

[19] Kawamoto H, Miwa T. Mitigation of lunar dust adhered to mechanical parts of equipment used for lunar exploration. Journal of Electrostatics. 2011;**69**:365

[20] Mazumder M, Horenstein M, Stark J, Hudelson JN, Sayyah A, Heiling C, et al. Proc. IAS Annu. Meet., 2014. 2014-EPC-0305

[21] Kawamoto H, Shibata T. Electrostatic cleaning system for removal of sand from solar panels. Journal of Electrostatics. 2015;**73**:65

[22] Roth JR, Sherman DM, Wilkinson SP. Boundary Layer Flow Control with a One Atmosphere Uniform Glow Discharge Surface Plasma. AIAA 98-0328. 1998

[23] Roth JR, Din X. Optimization of the aerodynamic plasma actuator as an electrohydro-dynamic (EHD) electrical device. AIAA, 44th AIAA Aerospace Sciences Meeting and Exhibit, 9-12 Jan., 2006

[24] Benard N, Moreau E. Electrical and mechanical characteristics of surface AC dielectric barrier discharge plasma actuators applied to airflow control. Experiments in Fluids. 2014;**55**:1846. DOI: 10.1007/s00348-014-1846-x

[25] Shimizu K, Mizuno Y, Blajan M, Yoneda H. Characteristics of an atmospheric nonthermal microplasma actuator. IEEE Transactions on Industry Applications. 2017;**53**(2):1452-1458

[26] Blajan M, Mizuno Y, Ito A, Shimizu K. Microplasma actuator for EHD induced flow. IEEE Transactions on Industry Applications. 2017;**53**(3):2409-2415

[27] Blajan M, Ito A, Kristof J, Shimizu K. Flow control by dielectric barrier discharge microplasma. In: Luca D, Sirghi L, Costin C, editors. Recent Advances in Technology Research and Education. INTER-ACADEMIA 2017. Advances in Intelligent Systems and Computing. Vol. 660. Cham: Springer; 2018

[28] Nguyen HP, Park MJ, Kim SB, Kim HJ, Baik LJ, Jo YM. Effective dielectric barrier discharge reactor operation for decomposition of volatile organic compounds. Journal of Cleaner Production. 2018;**198**:1232-1238

[29] Holzer F, Kopinke F-D, Roland U. Non-thermal plasma treatment for the elimination of odorous compounds from exhaust air from cooking processes. Chemical Engineering Journal. 2018;**334**:1988-1995

[30] Van Durme J, Dewulf J, Sysmans W, Leys C, Van Langenhove H. Abatement and degradation pathways of toluene in indoor air by positive corona discharge. Chemosphere. 2007;**68**:1821-1829

[31] Jung J-S, Kim J-G. An indoor air purification technology using a non-thermal plasma reactor with multiple-wire-to-wire type electrodes and a fiber air filter. Journal of Electrostatics. 2017;**86**:12-17

[32] Shimizu K, Muramatsu S, Kristof J, Blajan M. Analysis of hexadecane decomposition by atmospheric microplasma. IEEE Transactions on Industry Applications. 2018;**54**(1):605-610

[33] Shimizu K, Kuwabara T, Blajan M. Study on decomposition of indoor air contaminants by pulsed atmospheric microplasma. Sensors. 2012;**12**:14525-14536

[34] Shimizu K, Kurokawa Y, Blajan M. Study of VOC removal and *E. coli* sterilization in six-mat space by atmospheric microplasma. In: Proc. Int.

Conf. IEEE Ind. Appl. Soc., 2015-EPC-0512, Addison, TX, USA. 2015

[35] Nejadkoorki F. Advanced Air Pollution. Rijeka, Croatia: InTech; 2011. p. 584

[36] Shimizu K, Blajan M, Tatematsu S. Basic study of remote disinfection and sterilization effect by using atmospheric microplasma. IEEE Transactions on Industry Applications. 2012;**48**(4):1182-1188

[37] Blajan M, Shimizu K. Temporal evolution of dielectric barrier discharge microplasma. Applied Physics Letters. 2012;**101**:104101

[38] Aleknaviciute I. Plasma assisted decomposition of methane and propane and cracking of liquid hexadecane [Ph.D. dissertation]. Uxbridge, U.K.: School Eng. Design, Brunel Univ.; 2013. [Online]. Available from: http://bura.brunel.ac.uk/handle/2438/8574

[39] Kristof J, Miyamoto H, Tran AN, Blajan M, Shimizu K. Feasibility of transdermal delivery of cyclosporine a using plasma discharges. Biointerphases. 2017;**12**(2):02B402-1-02B402-11

[40] Shimizu K, Tran AN, Blajan M. Effect of microplasma irradiation on skin barrier function. Japanese Journal of Applied Physics. 2016;**55**(1):07LG01

[41] Kalghatgi S, Tsai C, Gray R, Pappas D. 22nd International Symposium on Plasma Chemistry. Antwerp, Belgium; 5-10 July 2015. p. O-22-6

[42] Szili EJ, Bradley JW, Short RD. A 'tissue model' to study the plasma delivery of reactive oxygen species. Journal of Physics D: Applied Physics. 2014;**47**:152002

[43] Kartaschew K, Mischo M, Baldus S, Brundermann E, Awakowicz P, Havenith M. Unraveling the interactions between cold atmospheric plasma and skin-components with vibrational microspectroscopy. Biointerphases. 2015;**10**:029516

[44] Marschewski M, Hirschberg J, Omairi T, Viol W, Emmert S, Maus-Friedrichs W. Electron spectroscopic analysis of the human lipid skin barrier: Cold plasma-induced changes in lipid composition. Experimental Dermatology. 2012;**21**:921

[45] Rai V, Ghosh I, Bose S, Silva SMC, Chandra P, Michniak-Kohn B. A transdermal review on permeation of drug formulations, modifier compounds and delivery methods. Journal of Drug Delivery Science and Technology. 2010;**20**:75-87

[46] Hill JR, Wertz PW. Molecular models of the intercellular lipid lamellae from epidermal stratum corneum. Biochimica et Biophysica Acta. 2003;**1616**:121. DOI: 10.1016/S0005-2736(03)00238-4

[47] Tachibana K. IEEJ Transactions on Electrical and Electronic Engineering. 2006;**1**:145

[48] Lauer JL, Shohet JL, Albrecht RM, Pratoomtong C, Murugesan R, Esnault S, et al. Journal of Applied Physics. 2004;**96**:4539

[49] Szili EJ, Al-Bataineh SA, Bryant PM, Short RD, Bradley JW, Steele DA. Plasma Processes and Polymers. 2011;**8**:38

[50] Mariotti D, Sankaran RM. Journal of Physics D. 2010;**43**:323001

[51] Melcher JR, Warren EP, Kotal RH. IEEE Transactions on Industry Applications. 1989;**25**:956

[52] Kawamoto H, Hasegawa N. Journal of Imaging Science and Technology. 2004;**48**:404

[53] Kern W. Journal of the Electrochemical Society. 1990;**137**:1887

[54] Zhang F, Busnaina AA, Fury MA, Wang S-Q. Journal of Electronic Materials. 2000;**29**:199

[55] Shimizu K, Ito A, Blajan M, Kristof J, Yoneda H. Basic study of fine particle removal using microplasma and its electrostatic effect. Japanese Journal of Applied Physics. 2017;**56**:01AC03. DOI: 10.7567/JJAP.56.01AC03

[56] Blajan M, Ito A, Kristof J, Shimizu K. Directional flow control with multielectrode system microplasma actuator. Journal of Biomedical Systems and Emerging Technologies. 2017;**4**:116

[57] Zhao L, Adamiak K. Electrohydrodynamic flow produced by electric corona discharge (numerical and experimental studies, and applications). In: Proc. 2009 Electrostatics Joint Conference. Vol. 13. Boston, America, P2; 2010. pp. 16-18

[58] Colas DF, Ferret A, Pai DZ, Lacoste DA, Laux CO. Ionic wind generation by a wire-cylinder-plate corona discharge in air at atmospheric pressure. Journal of Applied Physics. 2010;**108**:103306

[59] Robinson M. Movement of air in the electric wind of the corona discharge. Transactions of the American Institute of Electrical Engineers. 1961;**80**:143-150

[60] June MS, Kribs J, Lyons KM. Measuring efficiency of positive and negative ionic wind devices for comparison to fans and blowers. Journal of Electrostatics. 2011;**69**:345-350

[61] Kim C, Park D, Noh KC, Hwang J. Velocity and energy conversion efficiency characteristics of ionic wind generator in a multistage configuration. Journal of Electrostatics. 2010;**68**:36-41

[62] Touchard G. Plasma actuators for aeronautics applications–State of art review. International Journal of Plasma Environmental Science & Technology. 2008;**2**:1-25

[63] Whalley RD, Choi KS. The starting vortex in quiescent air induced by dielectric-barrier-discharge plasma. Journal of Fluid Mechanics. 2012;**703**:192-203

[64] Whalley RD, Choi KS. Turbulent boundary-layer control with plasma spanwise travelling waves. Experiments in Fluids. 2014;**55**:1796

[65] Choi KS, Jukes T, Whalley R. Turbulent boundary-layer control with plasma actuators. Philosophical Transactions of the Royal Society A. 2011;**369**:1443-1458

[66] Suzen YB, Huang PG, Jacob JD, Ashpis DE. Numerical simulations of plasma based flow control applications. In: 35th Fluid Dynamics Conference and Exhibit. Toronto, Ontario; June 6-9, 2005. AIAA 2005-4633

[67] Suzen YB, Huang PG, Ashpis DE. Numerical simulations of flow separation control in low-pressure turbines using plasma actuators. In: 45th AIAA Aerospace Sciences Meeting and Exhibit. Reno, Nevada; 8-11 January 2007. AIAA 2007-937

[68] Orlov DM. Modelling and Simulation of Single Dielectric Barrier Discharge Plasma Actuator [PhD Thesis]. University of Notre Dame; 2006

[69] Maden I, Maduta R, Kriegseis J, Jakirlic S, Schwarz C, Grundmann S, et al. Experimental and computational study of the flow induced by a plasma

actuator. International Journal of Heat and Fluid Flow. 2013;**41**:80-89

[70] Abdollahzadeh M, Pascoa JC, Oliveira PJ. Modified split-potential model for modeling the effect of DBD plasma actuators in high altitude flow control. Current Applied Physics. 2014;**14**:1160-1170

[71] Brauner T, Laizet S, Benard N, Moreau E. Modelling of dielectric barrier discharge plasma actuators for direct numerical simulations. In: AIAA Aviation. Washington, D.C.: 8th AIAA Flow Control Conference; 13-17 June 2016. AIAA 2016-3774

Chapter 5

Atmospheric Plasma Spray Processes: From Micro to Nanostructures

Felipe Miranda, Felipe Caliari, Alexei Essiptchouk and Gilberto Pertraconi

Abstract

Atmospheric plasma spray is probably the most versatile of all thermal spraying processes, because there are few limitations either on the materials that can be sprayed or the substrate, in relation to its material, size, and shape. The material precursor of the coating could be in the form of powders, wires, melted materials, solutions, or suspensions. What distinguishes the plasma spray process from other technologies is its applicability and capacity to process a wide variety of materials, including metallic and refractory materials at atmospheric pressure. The coatings properties are improved by deposition of coatings with finer microstructure, which is are more suitable for mechanical and thermal stresses than the lamellar microstructure of conventional plasma-sprayed coatings.

Keywords: atmospheric plasma spray, solution plasma spray, thermal barrier coatings, environmental barrier coatings, nanostructured coating

1. Introduction

Atmospheric plasma spray (APS) appeared after Second World War as a surface finishing technology. It is now widely used to deposit thick coatings (from hundreds of micrometers up to a few millimeters) in a substrate to protect in aggressive environments or to improve its function. APS is commonly used in many industrial sectors, including aeronautics, energy, automotive, mining, biomedical, and electronics [1]. The synthesis of coatings by APS technique occurs by stacking the lamellae resulting from the impact, flattening, and solidification by the colliding molten particles [2]. The material precursor of the coating may be in the form of powders, wires, melt materials, solutions, or suspensions [3]. APS can be applied to a wide variety of materials, including metallic and refractory materials. In this technique, a carrier gas conducts the material particles by injecting them at high velocity through the plasma, where they are molten or partially molten, taking the form of droplets that settle and solidify on the surface being coated. The material to be deposited is carried in the form of a solution or powder to a torch with sufficient enthalpy to generate a plasma jet to melt the particles [4, 5]. The parameters of the plasma spraying process, as well as the characteristics of the precursor (solids or liquids) used for coating, influence the properties of the deposited materials [6, 7]. Characteristics of the coatings such as porosity, atomic structure, roughness, cohesion, and adhesion are fundamentally related to the interaction of the precursor with the plasma jet [8, 9].

The main driving force for the manufacture of thick coatings by APS is their high deposition rate; a few kilograms per hour of raw material can be processed with torches, with a power level of a few tens of kilowatts at a relatively low operating cost. Plasma spray is probably the most versatile of all thermal spray processes because there are few limitations of materials that can be sprayed or on the material, size, and shape of the substrate [4]. The coatings are characterized by a highly anisotropic lamellar structure. In addition, stacking of the particles generates specific interlamellar characteristics throughout the structure, especially voids, which may connect or not through the particles encountering the coating thereafter.

Conventional plasma spray processes (CPS) use powders with a particle size ranging from 10 to 100 μm. Typically, these result in coatings formed by lamellae of micrometer thickness and diameter of a few tens to hundreds of micrometers. The interest in developing and studying plasma spray coatings that have nanometric and non-micrometric characteristics has been the focus of the last 30 years. This interest stems from improved nanometric coating properties compared to micron size [10]. Reducing the structure to the nanoscale improves hardness, elasticity modulus, and thermal conductivity of the coating as well as reduces defects (voids). One of the main drawbacks in the processing of nanometric particles by APS is the difficulty in injecting them into the high enthalpy jet core. In this case, it is necessary to adjust the injection angle and the transport gas flow of the material in such a way that it is not so intense that it crosses the plasma jet or so smooth that it cannot reach the center of the jet [11]. For this reason, new plasma torches have been developed using axial injection, which is a method by which the material is injected directly into the plasma torch, ensuring that all material is processed. This feature assists in the processing of materials with nanometric magnitude, which can mainly be dispersed in liquid medium to facilitate loading and processing [12].

2. Why apply coatings?

Coatings are applied to substrates (metallic, ceramic, polymeric, or composite) to incorporate to their surface one or more characteristics or qualities that they do not originally possess, to maximize the useful life of a material/equipment, by increasing its resistance to corrosion, wear, oxidation, thermal protection, or gaining efficiency, depending on its application [13].

The improved performance of a given component means, in addition to a providing greater longevity, reduces costs related to the project and subsequent maintenance. In this context, the choice of material is as important as the design of the final product. New materials have been developed for the most diverse applications, which are subjected to extreme conditions, such as in high temperatures, abrasive wear, oxidizing and corrosive atmospheres, or even in a combination of these conditions. However, it is impossible for a single material to meet all these requirements, even for special alloys. This fact contributes to increased demand for the applied coatings, which can be deposited in substrates of most materials and with properties adjusted to specific necessities [14].

Coatings can be divided into thin and thick films. The thin films have thicknesses up to 20 μm and are typically synthesized by chemical vapor deposition (CVD) or physical vapor deposition (PVD). However, most thin film deposition processes require reactors that work at low pressures, increasing process costs and limiting the dimensions and geometry of the substrate to be coated. Another category of coatings are thick films, which have thicknesses greater than 20 μm, or even a few millimeters. These are applied when the performance of the protection depends directly of the thickness of the coating, for example, coatings applied in aerospace devices, which are subjected to severe erosion, corrosion, and oxidation environments. The deposition methods of thick films include chemical/electro-chemical plating, brazing, solder overlays, and plasma spray that are performed at atmospheric pressure [15].

The evolution of these coatings is often the limiting factor in the development of a technology. For example, the input temperature of the first gas turbine (1933) was 400°C. In 2011, the Japanese company Mitsubishi Hitachi Power Systems developed an advanced gas turbine with input temperature of 1600°C, and efforts are under way to raise this temperature to 1700°C [16]. Initially this evolution occurred through the development of super leagues and the consolidation of these by directional solidification and monocrystals. The construction projects also promoted advances, mainly in the cooling through internal channels of the parts in the hot section of the turbine, where cold air is injected. However, the greatest advance in the temperature operation of these thermal machines occurred with the application of thermal barrier coatings (TBC), which are applied with APS process [17]. The TBCs (**Figure 1**) are composed of three layers: the first layer is a metallic layer (bond coat), which provides resistance to oxidation and corrosion, and has an intermediate coefficient of thermal expansion between the substrate and the ceramic layer. The second is a layer of oxide (thermal grown oxide (TGO)) that appears, along the thermal cycles, in the interface between the metallic layer and ceramic, due to the oxidation of the metallic layer. The third layer is a ceramic (top coating), which is responsible for the durability of TBC as a whole, as it promotes thermal insulation and protects the metallic layer and substrate from exposure to hot and oxidizing gases from combustion.

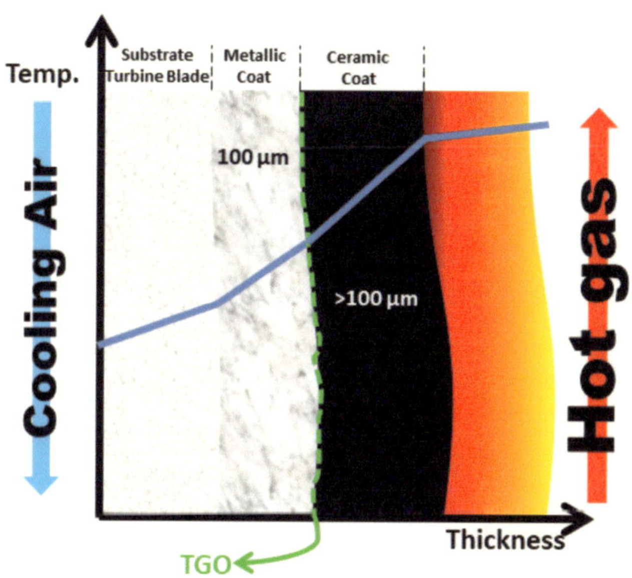

Figure 1.
Thermal barrier coatings applied on turbine blade.

3. Atmospheric plasma spray process

Atmospheric plasma spray belongs to a family of processes called thermal spray. Thus, the thermal spray process encompasses a set of processes where sprayed materials (powder on the order of μm or solutions in the form of suspensions, solutions, sol-gel or colloids) are heated (to be melted or semi-melted or partially evaporated) and accelerated against a substrate to form a coating, which may be lamellar or columnar. The thermal source may be by combustion of hydrocarbons or electric arc. The plasma spray process consists of forming the plasma jet, which interact the particles with the plasma. The plasma jet provides thermal and kinetic energy to the material to be deposited by directing it to a substrate to produce coating [18].

In the deposition process of plasma spray, the substrate is usually prepared to receive the coating by performing cleaning procedures (removing oils and greases), inducing surface roughness, preheating, and controlling movement. The adhesion of the particles to the substrate and between the lamellae strongly depends on the preparation of the substrate, preheating temperature, and morphology and composition of the material. Preheating, generally performed with the plasma jet itself, is a key point and has to be controlled according to the size and thickness of the part to be coated. Substrate and coating temperatures, either for preheating or during deposition, are linked to the residual voltage distribution, which is a controlled parameter [19].

Another important step in the deposition process is the material injection, which is mainly influenced by the injection method, feed system type, and the characteristics of the material. The injection method may be radial or axial (**Figure 2**).

In the radial injection method, the parameters to be optimized primarily include the flow of the loading gas and the position and geometry of the injector. The point is to ensure that the momentum density of the particles (product of the specific mass by velocity) is equal to the momentum of the plasma jet at the point of injection. The radial injection method has restrictions and some negative aspects, such as the heterogeneity of the heating and the acceleration of the particles, according to their granulometric distribution; and the difficulty of processing precursors with high particle velocity, because of the low plasma-particle heat transfer efficiency. The APS process should attempt to maximize the residence time of the particles in the plasma jet. To overcome the residence time problem, small amounts of hydrogen are added to the working gas to increase the enthalpy of the plasma jet; however, this also increases the electric arc oscillation [20].

Another method is axial injection, which is used in flame spray, high velocity oxygen fuel (HVOF), induction (RF) plasma torches, and some plasma torches from a direct current source (DC). In axial injection the distribution of particles are keeping more concentrated, and the interaction time between plasma-particle is

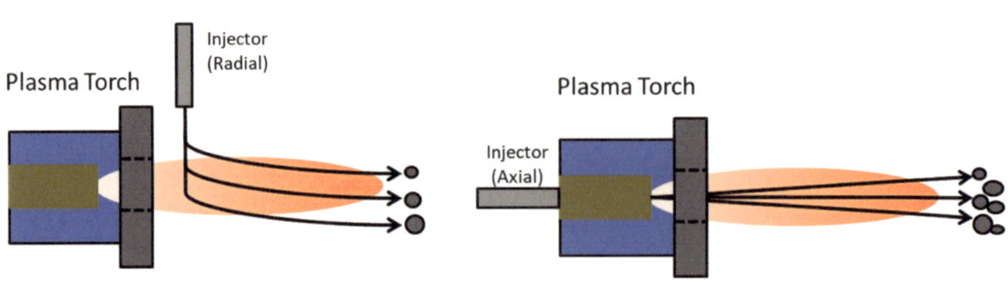

Figure 2.
Radial and axial material injection.

greater, which ensures a better processing of the material. With higher heat transfer efficiency between plasma/particle, there is a lower incidence of unmelted particles that may weaken the coating. The problems are the clogging of the nozzle and the operational stability of the plasma torch. Thus, there is a need to develop new plasma torches that operate better with this injection method.

Caliari et al. [21] presented a plasma torch with axial material injection, which can produce high velocity plasma jet (>1200 m/s). The materials are injected into the back of the plasma torch (**Figure 3**), ensuring that all material passes through hot area near of the electric arc, which is the hottest part of the torch, thereby obtaining a lower rate of unmelted particles. The process called "High Velocity Plasma Spray" has a stable torch that operates with low currents between 50~150 A and voltage of 240~360 V, unlike conventional torches, which operate at low voltages and up to five times higher current values. This differential guarantees lower erosion of the electrodes, and its geometry provides better use of the energy with close to 80% efficiency of electric to thermal energy conversion. This high efficiency enables the processing of metallic and ceramic materials, both solids or liquids precursors (suspensions or solutions) [12].

3.1 Plasma spray system

The principal components of the plasma spray systems are (see **Figure 4**): plasma torch, process control, power supply, gas supply, material feeder (powder or liquid), and dynamic sample holder. The main element of a plasma spray system is the plasma torch, which is responsible for converting electric energy into thermal energy, necessary for material processing. A process control console

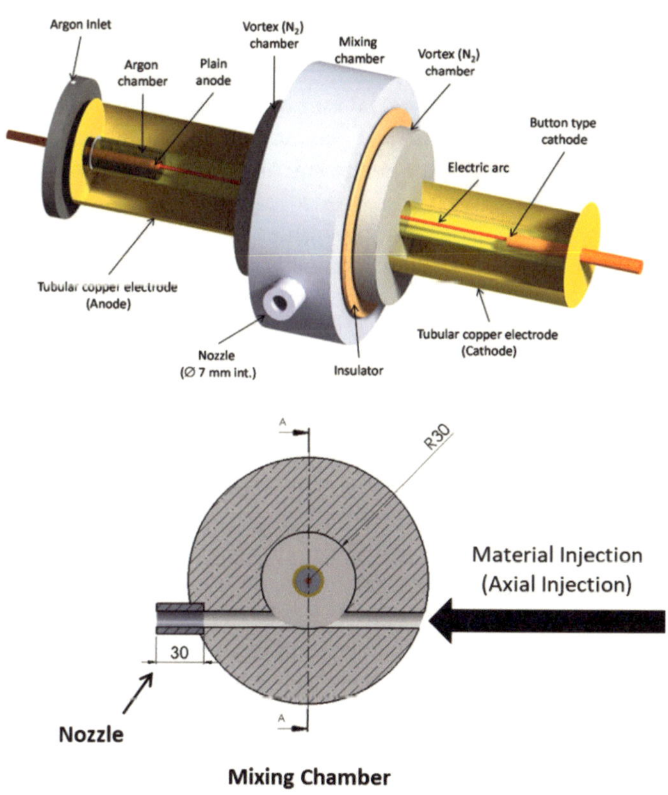

Figure 3.
High velocity plasma spray with axial injection presented by Caliari et al. [21].

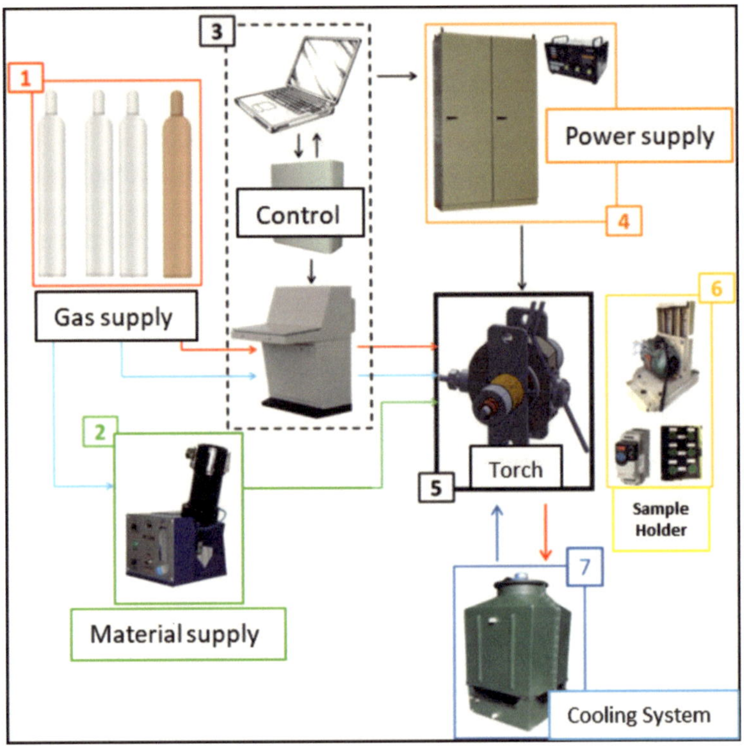

Figure 4.
Plasma spray system: (1) gas supply, (2) material supply, (3) control panel, (4) DC power supply, (5) plasma torch, (6) sample holder, and (7) cooling system.

allows adjustment of the operating parameters, i.e., the control of arc current, arc ignition, plasma gas flow rates, material and carrier gas flow rates. The additional systems necessary for operation are the plasma gas supply system, the power supply system (including the high-frequency starter unit), the high-pressure cooling water system, and the material (powder or solutions) feed system. The systems also have mechanical equipment (sample holder) to control relative motion between the plasma torch and the substrate.

3.2 Plasma torch: general aspect

Plasma torches are devices used to stabilize an electric discharge with gas flow and to convert electric energy into thermal energy. In a thermal plasma torch, operating from an electric discharge, the high enthalpy plasma results from the interaction of the gas with the electric arc. The study of electric discharges in gases and plasma jet formation involves the phenomena of gas dynamics, mass and heat transfer, and electrophysical and aero-thermodynamic processes [22].

Plasma torches can be classified according to the source of electrical energy (electric arcs generated from a direct current, DC, or alternating current, AC source) or by the type of discharge used (transferred or non-transferred arc). Transferred arc torches have one of the electrodes external to the torch body, through which the arc extends from the inner electrode. Due to the electric current transport in the generated plasma jet, this configuration forms higher enthalpy plasma jets than non-transferred arc torches. For non-transferred arc torches, both electrodes are positioned inside the torch. Thus, the electric arc remains confined in the discharge channel and the generated plasma jet does not carry electric current [23].

In the design of a thermal plasma torch, one should consider the type of electric power source, the enthalpy and temperature of the plasma jet suitable for the application, the choice of appropriate materials, the implementation of arc stabilization, and control system for the length of the electric arc (if any). In the case of non-transferred arc torches, the stabilization of the electric arc can be done with a gas vortex (that forms the plasma), discharge chamber wall, magnetic field and its combinations. In most cases the arc self-fixation method, based on shunting effect, is used. In addition, to fix the arc length, a magnetic field generated by one or more solenoids may be used. In this case, the radial part of the electric arc moves axially to the place where the axial component of the magnetic field is greater. Moreover, the interaction between the magnetic field and the electric arc current produces the driving force that displaces the arc tangentially, which avoids the positioning of the arc spot at a fixed point, thus reducing erosion of the electrode [24].

The materials applied in plasma torches, especially those exposed to the electric arc, are submitted to a high thermal load in the place of fixation of the electric arc, that destroyed the electrode. Properties such as specific heat, melt temperature, coefficient of thermal expansion, thermal conductivity, work function, and electrical resistivity should be considered in the choice of these materials.

Figure 5 presents a generic scheme of a non-transferred arc plasma torch with the electrodes (cathode, 1, and anode, 2) arranged concentrically and between the electric arc [25]. To stabilize the arc, the gas vortex formed by the vortex camera (4) installed between the insulated electrodes (3). To fix the arc in the anode surface, a magnetic field produced by the solenoids (7) is applied. The electrodes are subjected to high heat flows, and to maintain their functionality, they are cooled by a continuous flow of water through the cooling jacket (6).

Although there is a great diversity of plasma torch designs, their principle of operation is based on the generation of plasma flow due to forced convective interaction of a gas with the electric arc, established between two electrodes [23]. The application of a high frequency and high voltage at electrodes allows transform the electrical insulator gas into conductor and form an electric conductor channel. The high temperature in the channel, caused by Joule effect reduces the electrical resistance of the gas, due to an increase in the number of charged particles, thus allowing the passage of high current by the gas, establishing the arc. The electrons, accelerated by the electric field in the region between the electrodes, transfer their kinetic energy to the heavy particles through collisions, raising the temperature of the gas, dissociating its molecules, and exciting and ionizing the atoms, which are factors that contribute to the increased degree of plasma ionization. The electric field generated near the cathode accelerates the

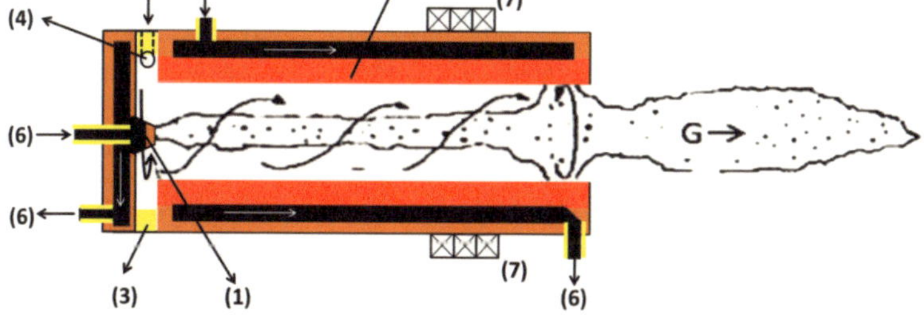

Figure 5.
Diagram of a linear thermal plasma torch, (1) hot cathode, (2) anode, (3) electrical insulation, (4) vortex chamber, (5) gas inlet, (6) cooling water inlet and outlet, and (7) solenoid.

positively charged heavy particles, which collide against the surface of the cathode. As the mobility of heavy particles is much smaller than that of the electrons, an excess of positive volume of charged particles is formed in the region near the cathode. This phenomenon increases the electric field in the vicinity of the electrode, which in turn, facilitates the emission of the electrons of the electrode (due to the tunneling and field effect) by increasing the density of electrons in the plasma [23]. The constriction of the arc in the cathode (another important phenomenon) increases the current density and, respectively, the thermal flow to the surface of the cathode, increasing its temperature and the emission of the electrons by thermionic effect.

4. Main parameters of the atmospheric plasma spray process

The input parameters and the operational characteristics of the plasma spray process are described in **Table 1** [15]. The input parameters are controlled during the experiment and, therefore, are independent variables. However, the operational characteristics often depend on the combination of the input parameters, thus, they are dependent variables.

To obtain operational control of the plasma torch, it is necessary to know the operating range of the input parameters in order to stabilize the electric arc. Hence, one must know the current-voltage characteristics (SVS) (voltage—versus current for different flow rates) of the electric arc, (CVC), which usually influenced by the characteristic curve of the electric power source. According to Heimann [26], the most common gases used in plasma torches are nitrogen, argon, helium, and hydrogen. Gas flow and its chemical composition, current and outlet electrode (nozzle) design directly influence the energetic and kinetic characteristics of the plasma jet and its stability. The arc voltage, in turn, depends on the gas flow rate, plasma torch geometry and mode of arc stabilization. Thermal efficiency represents the ability to convert electrical energy into thermal one (enthalpy of the plasma jet). Part of the thermal energy is dissipated in the electrodes with cooled walls. In conventional plasma torches, the thermal efficiency is approximately 50% [15].

	Torch	Plasma Jet	Particles	Substrate
Input parameters	• Current • Plasma gas composition • Flow rate • Nozzle deign, erosion • Cooling water flow		• Size distribution • Morphology • Feed rate • Carrier gas flow rate	• Substrate material • Pretreatment • Motion • Distance
Operating characteristics	• Voltage • Voltage fluctuations • Thermal efficiency • Torch set up	• Stability • Geometry • Plasma gas composition • Plasma jet temperature and velocity • Atmosphere • Pressure Humidity	• Particle trajectory • Particle temperature • Particle velocity • Particle distribution	• Coatings properties • Porosity • Mechanical properties • Deposition efficiency

Table 1.
The main parameters and characteristics of the APS process [15].

5. Atmospheric plasma spray micro-to-nano-sized structures

The applications of atmospheric plasma spray technology have changed considerably since its beginning in the 1950s. The global pressures on prices have forced companies to face challenges in their manufacturing processes; they generally answer by an acceleration of production, increasing in throughput and consistency in quality of the coating. Also, plasma-sprayed coatings have to cover a greater demand of applications such as higher operation temperatures, wear and corrosion under extreme conditions, and longer life span of parts and devices. A potential response for coatings with improved properties is the deposition of coatings with finer microstructure, i.e., finer lamellae and smaller voids as well as coating with microstructure more resistant to mechanical and thermal stresses than the lamellar microstructure exhibited by conventional plasma-sprayed coatings. This requirement has led to the development of innovative plasma coating processes for producing coatings with grain size in the nanometer range while keeping the high deposition rate and flexibility of plasma spraying [27, 28]. The process uses the basic equipment of the conventional plasma spray process but the feedstock is a liquid suspension or a solution of chemical precursors instead of the conventional powder feedstock, which takes advantage of the high enthalpy content of the plasma jet to evaporate the spray material and then forms a coating by fine droplets and/or condensation of the vaporized material on the substrate [29, 30]. To form the coating with liquid precursors, these must be injected into the plasma region in smaller droplets (sprayed or atomized). Then the solvent is evaporated, forming the solid material, which is melted (or forms a shell) and accelerated towards the substrate, as shown in **Figure 6**. As all these steps occur almost instantaneously, the plasma generator must provide suitable energy to process the liquid precursor and form the material.

Available since the 1990s, nanostructured materials are still considered a new concept that increase the performance of engineering components. Many studies have instigated the properties of nanostructured materials used in structural components and coatings. Ceramic materials gained attention mainly because they have greater hardness due to the smaller grain size, greater resistance to wear, and less incidence of defects [10, 31].

Nanostructured coatings can be obtained by plasma spray processes using liquid precursors. Solid precursors (powders) with nanometric distribution have difficulty with fluidity between the feed line and the plasma torch, causing intermittent injection of the material. The fluidity problem is overcome by increasing the flow of the carrier gas; however, this causes the plasma jet axis to shift and produce a non-uniform coating on the substrate. Another factor is that the nanoparticle may not penetrate the center of the plasma jet by inhibiting fusion and acceleration processes towards the substrate to form the coating [5, 32–35].

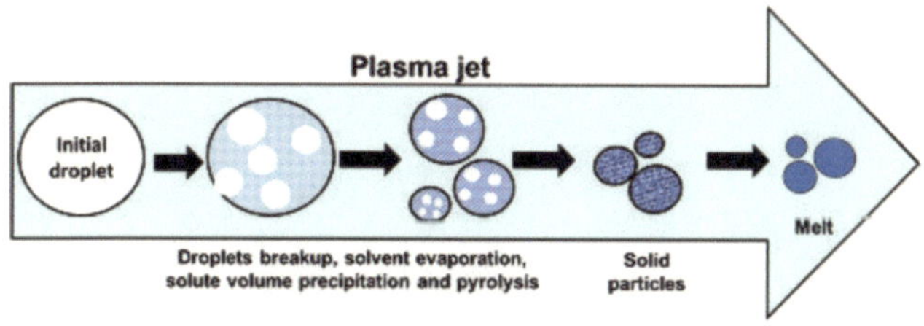

Figure 6.
Formation of the material from a liquid precursor with low and high concentration of the material to be deposited.

The solution plasma spray technique allows the deposition of thick nanostructured coatings (**Figure 7**), without the need for a very sophisticated infrastructure. Expenditure on materials (liquid precursor) is much lower than on the powders. The flexibility to use different feedstock enables a variety of compositions of the liquid precursor to be exploited by adjusting its concentration according to the desired application.

Biomaterials can be obtained with different techniques, thermal spraying shows significant advantages; in particular, the fact that the deposition and consolidation of the coating occur simultaneously without the need of a sintering treatment. Bioactive glasses are considered promising materials to be used as coatings onto implant devices, due to their high bioactivity [36, 37]. The use of solutions precursor instead of traditional thermal spraying feedstock provides unique properties, i.e., high purity materials (avoiding possible contamination from feedstock preparation steps), and nanostructured coatings with denser and more homogeneous microstructures [38].

Several efforts to use solution plasma spray process to fabricate superhydrophobic coatings have been reported [39]. Metals and metal oxides, as the most important and commonly used engineering materials, are hydrophilic for most part due to their high surface energy. There is immense interest in developing the ability to control the surface wettability of metals and metal oxides in order to improve their performance in corrosion resistance, friction reduction and efficiency in liquid transportation [40]. Xu et al. [41] presented superhydrophobic ceramic coatings with nano-sized hierarchical structure and high water contact angle, coatings were fabricated by a one-step solution precursor plasma spray process.

In addition, plasma reactivity can be exploited to obtain a final coating with composition different from the original precursor. In the study presented by Miranda et al. [12], a plasma spray system with axial injection was used to deposit nanostructured coatings. Coatings with graded composition between SiO_2/SiC (**Figure 8**) on carbon/carbon composites substrates were obtained.

A relevant result, obtained during the analysis of the chemical and structural composition of the coating, was the SiC formation due to the reactions between the carbon and the liquid precursor (SiO) promoted by the plasma jet. These reactions are exemplified by means of **Figure 9**. The formation of SiC in the coating helps to protect the substrate because it reduces the permeability of oxygen, preventing its oxidation and, consequently, the loss of its structural characteristics. Although no coating adhesion tests were carried out, the higher SiC concentration at the substrate/coating interface indicates the occurrence of a chemical adhesion process of the coating [42]. Thus, at this stage of the deposition process, the composite is

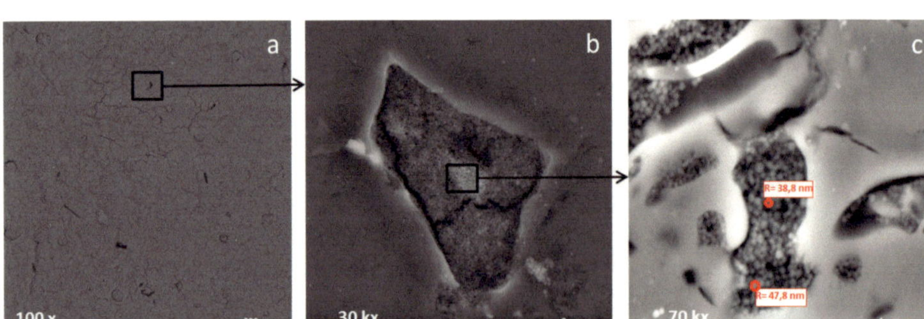

Figure 7.
(a) Top view of the coating, (b) nanostructure of coatings, and (c) particle measurements.

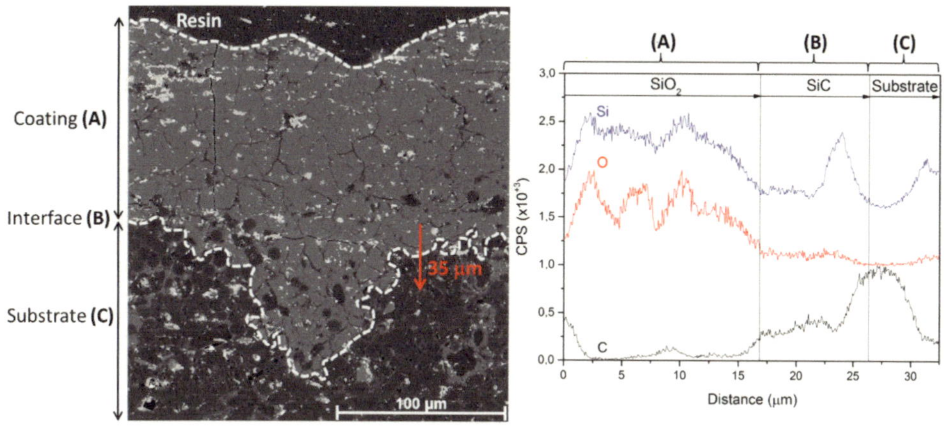

Figure 8.
SEM image of cross-section of sample and EDS results showing the composition of coating (A), interface (B), and substrate (C).

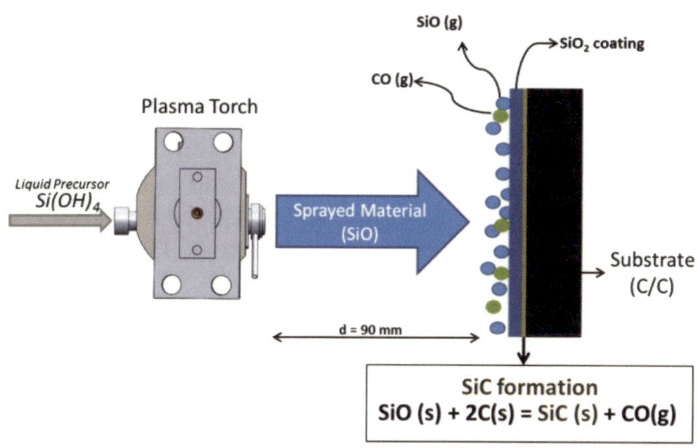

Figure 9.
SiC formation in the deposition process [12].

"doped" with a SiC layer and thus exposed to differentiated oxidation processes in relation to the original C/C substrate.

The coatings aim at the environmental protection of carbon/carbon composite substrates of great application in the aerospace sector. The main tool of this system is the non-transferred arc plasma torch calibrated and specifically characterized by forming a high enthalpy and high velocity plasma jet, capable to processing precursors with high melt temperatures at atmospheric pressure. With these characteristics, adjusted to power systems, gas feed and liquid precursors, it is possible to obtain nanostructured coatings with low fraction of pores and inclusions.

In processes using DC or RF plasma torches, typically the material is introduced into the plasma jet in the form of micrometric powders. They are accelerated due to moment transfer of the plasma jet and at the same time the process of heat transfer and mass begins. The residence time of the powder in the plasma jet, and consequently the efficiency of the process, depend on the particle's speed and the flight distance. The process of forming nanostructured coatings relates to these factors, since particles within the plasma flame must remain a sufficient time for total evaporation of the liquid precursor and melt the particles or particle's surface. The formation of nanostructures occurs with the rapid cooling of the vapor.

It is important to point out that this process is most commonly observed in works that employ short-arc plasma torches transferred for better utilization of process energy [43, 44]. In these systems, the vapor cooling process is the main stage of nanoparticle formation. Homogeneous nucleation is facilitated by the combination of heating, evaporation, and rapid cooling, such as happens in a plasma reactor [45].

6. Conclusions

Nanostructured materials have unique characteristics in relation to their mechanical and thermal properties, in addition to producing a lower index of defects and a lower porosity than coatings obtained in the traditional method using solid feedstock (which are limited to the size of the particle in flight). The use of atmospheric spray plasma systems is shown as an effective alternative to produce nanostructured and composite materials, which can be adjusted according to the need for application, as in the case of the use of liquid precursors. Therefore, it is necessary to advance the development of plasma torches capable of processing various types of materials to increase its range of application and consequently contribute to the technological advancement of materials processing.

Acknowledgements

The authors acknowledge the financial support grant #88887.185537/2018-00 provided by the Coordination for the Improvement of Higher Education Personnel (CAPES) and the Technological Institute of Aeronautics (ITA).

Author details

Felipe Miranda[1*], Felipe Caliari[1], Alexei Essiptchouk[2] and Gilberto Pertraconi[1]

1 Technological Institute of Aeronautics (ITA), São José dos Campos, Brazil

2 São Paulo State University (UNESP), São José dos Campos, Brazil

*Address all correspondence to: mirannda.fs@gmail.com

References

[1] Davis, Joseph R. ASM Handbook, Volume 5A: Thermal Spray Technology. In: Robert C, editor. Materials Park, OH, USA: ASM International; 2013

[2] Davis JR. Handbook of Thermal Spray Technology. Materials Park, OH: ASM International; 2004

[3] Hermanek FJ. Thermal spray terminology and company origins. Materials Park, OH: ASM International; 2001

[4] Fauchais P. Understanding plasma spraying. Journal of Physics D: Applied Physics. 2004;**37**:R86-R108. DOI: 10.1088/0022-3727/37/9/R02

[5] Xie L, Ma X, Jordan EH, Padture NP, Xiao DT, Gell M. Deposition mechanisms of thermal barrier coatings in the solution precursor plasma spray process. Surface and Coatings Technology. 2004;**177-178**:103-107. DOI: 10.1016/j.surfcoat.2003.06.013

[6] Sampath S, Jiang X. Splat formation and microstructure development during plasma spraying: Deposition temperature effects. Materials Science and Engineering A. 2001;**304-306**:144-150. DOI: 10.1016/S0921-5093(00)01464-7

[7] Cao XQ, Vassen R, Stoever D. Ceramic materials for thermal barrier coatings. Journal of the European Ceramic Society. 2004;**24**:1-10. DOI: 10.1016/S0955-2219(03)00129-8

[8] Guo FA, Trannoy N, Gerday D. An application of scanning thermal microscopy: Analysis of the thermal properties of plasma-sprayed yttria-stabilized zirconia thermal barrier coating. Journal of the European Ceramic Society. 2005;**25**:1159-1166. DOI: 10.1016/j.jeurceramsoc.2004.04.026

[9] Xie L, Chen D, Jordan EH, Ozturk A, Wu F, Ma X, Cetegen BM, Gell M. Formation of vertical cracks in solution-precursor plasma-sprayed thermal barrier coatings. Surface and Coatings Technology. 2006;**201**:1058-1064. DOI: 10.1016/j.surfcoat.2006.01.020

[10] Gell M. Application opportunities for nanostructured materials and coatings. Materials Science and Engineering A. 1995;**204**:246-251. DOI: 10.1016/0921-5093(95)09969-7

[11] Fauchais P, Etchart-Salas R, Rat V, Coudert JF, Caron N, Wittmann-Ténèze K. Parameters controlling liquid plasma spraying: Solutions, sols, or suspensions. Journal of Thermal Spray Technology. 2008;**17**:31-59. DOI: 10.1007/s11666-007-9152-2

[12] Miranda FS, Caliari FR, Campos TM, Essiptchouk AM, Filho GP. Deposition of graded SiO_2/SiC coatings using high-velocity solution plasma spray. Ceramics International. 2017;**43**:16416-16423. DOI: 10.1016/j.ceramint.2017.09.018

[13] N.R. Council. Coatings for High-Temperature Structural Materials: Trends and Opportunities. Washington, DC: The National Academies Press; 1996. DOI: 10.17226/5038

[14] Fauchais P, Vardelle A, Vardelle M. Current Knowledge and Challenges in Suspensions Spraying. In: 4th International Round Table on Thermal Plasmas for Industrial Applications, March 3-7, 2014. Marrakesh , MA; 2014

[15] Fauchais PL, Heberlein JVR, Boulos MI. Thermal Spray Fundamentals: From Powder to Part. New York, USA: Springer; 2014

[16] Breeze P. Gas-Turbine Power Generation. 1st ed. Academic Press, 2016. 104 p. DOI: 10.1016/B978-0-12-804005-8.00007-0

[17] Perepezko JH. The hotter the engine, the better. Science. 2009;**326**(80):1068-1069. DOI: 10.1126/science.1179327

[18] Fauchais P, Montavon G, Lima RS, Marple BR. Engineering a new class of thermal spray nano-based microstructures from agglomerated nanostructured particles, suspensions and solutions: An invited review. Journal of Physics D: Applied Physics. 2011;**44**:093001. DOI: 10.1088/0022-3727/44/9/093001

[19] Fauchais P, Vardelle A, Vardelle M, Fukumoto M. Knowledge concerning splat formation: An invited review. Journal of Thermal Spray Technology. 2004;**13**:337-360. DOI: 10.1361/10599630419670

[20] Fauchais P, Vardelle M, Vardelle A. Reliability of plasma-sprayed coatings: Monitoring the plasma spray process and improving the quality of coatings. Journal of Physics D: Applied Physics. 2013;**46**:224016-224032. DOI: 10.1088/0022-3727/46/22/224016

[21] Caliari FR, Miranda FS, Reis DAP, Filho GP, Charakhovski LI, Essiptchouk A. Plasma torch for supersonic plasma spray at atmospheric pressure. Journal of Materials Processing Technology. 2016;**237**:351-360. DOI: 10.1016/j.jmatprotec.2016.06.027

[22] Solonenko O. Thermal plasma torches and technologies: Plasma torches, basic studies and design. Vol. 1. Annals of Physics, New York. 2003:397

[23] Zhukov MF, Zasypkin IM, Timoshevskii AN, Mikhailov BI, Desyatkov GA. Thermal Plasma Torches: Design, Characteristics, Application. Cambridge: Cambridge International Science Publishing; 2007

[24] Essiptchouk AM, Charakhovski LI, Filho GP, Maciel HS, Otani C, Barros EA. Thermal and power characteristics of plasma torch with reverse vortex. Journal of Physics D: Applied Physics. 2009;**42**:175205. DOI: 10.1088/0022-3727/42/17/175205

[25] Marquesi AR. Estudo e Desenvolvimento de Uma Tocha de Plasma de Vapor de Água Para Aplicações em Processos de Gaseificação. Instituto Tecnológico de Aeronáutica; 2016

[26] Heimann RB. Plasma-Spray Coating: Principles and Applications. New York: VCH Publisher, Inc.; 1996. 340 p. DOI: 10.1002/9783527614851

[27] Karthikeyan J, Berndt CC, Tikkanen J, Reddy S, Herman H. Plasma spray synthesis of nanomaterial powders and deposits. Materials Science and Engineering A. 1997;**238**:275-286. DOI: 10.1016/S0921-5093(96)10568-2

[28] Bouyer E, Gitzhofer F, Boulos MI. The suspension plasma spraying of bioceramics by induction plasma. Journal of Metals. 1997;**49**:58-62. DOI: 10.1007/BF02915483

[29] Smith MF, Hall AC, Fleetwood JD, Meyer P. Very low pressure plasma spray—A review of an emerging technology in the thermal spray community. Coatings. 2011;**1**:117-132. DOI: 10.3390/coatings1020117

[30] Von Niessen K, Gindrat M. Plasma spray-PVD: A new thermal spray process to deposit out of the vapor phase. Journal of Thermal Spray Technology. 2011;**20**:736-743. DOI: 10.1007/s11666-011-9654-9

[31] Fauchais P, Vardelle M, Goutier S, Vardelle A. Key challenges and opportunities in suspension and solution plasma spraying. Plasma Chemistry and Plasma Processing. 2015;**35**:511-525. DOI: 10.1007/s11090-014-9594-5

[32] Delbos C, Fazilleau J, Rat V, Coudert JF, Fauchais P, Pateyron B.

Phenomena involved in suspension plasma spraying part 2: Zirconia particle treatment and coating formation. Plasma Chemistry and Plasma Processing. 2006;**26**:393-414. DOI: 10.1007/s11090-006-9020-8

[33] Viswanathan V, Laha T, Balani K, Agarwal A, Seal S. Challenges and advances in nanocomposite processing techniques. Materials Science & Engineering R: Reports. 2006;**54**:121-285. DOI: 10.1016/j.mser.2006.11.002

[34] Gell M, Jordan EH, Teicholz M, Cetegen BM, Padture NP, Xie L, Chen D, Ma X, Roth J. Thermal barrier coatings made by the solution precursor plasma spray process. Journal of Thermal Spray Technology. 2008;**17**:124-135. DOI: 10.1007/s11666-007-9141-5

[35] Pawlowski L. Suspension and solution thermal spray coatings. Surface and Coatings Technology. 2009;**203**:2807-2829. DOI: 10.1016/j.surfcoat.2009.03.005

[36] Sola A, Bellucci D, Cannillo V, Cattini A. Bioactive glass coatings: A review. Surface Engineering. 2011;**27**:560-572. DOI: 10.1179/1743294410Y.0000000008

[37] Baino F, Verne E. Glass-based coatings on biomedical implants: A state-of-the-art review. Biomedical Glasses. 2017;**3**(1):1-17

[38] Killinger A, Gadow R, Mauer G, Guignard A, Vaßen R, Stöver D. Review of new developments in suspension and solution precursor thermal spray processes. Journal of Thermal Spray Technology. 2011;**20**:677. DOI: 10.1007/s11666-011-9639-8

[39] Cai Y, Coyle TW, Azimi G, Mostaghimi J. Superhydrophobic Ceramic Coatings by Solution Precursor Plasma Spray. Scientific Reports. 2016;**6**:24670

[40] Liu K, Jiang L. Metallic surfaces with special wettability. Nanoscale. 2011;**3**:825-838. DOI: 10.1039/C0NR00642D

[41] Xu P, Pershin L, Mostaghimi J, Coyle TW. Efficient one-step fabrication of ceramic superhydrophobic coatings by solution precursor plasma spray. Materials Letters. 2018;**211**:24-27. DOI: 10.1016/j.matlet.2017.09.077

[42] Stern KH. Metallurgical and Ceramic Protective Coatings. 1st ed. Washington, DC: Chapman & Hall; 1996

[43] Pfender E, Lee YC. Particle dynamics and particle heat and mass transfer in thermal plasmas. Part I. The motion of a single particle without thermal effects. Plasma Chemistry and Plasma Processing. 1985;**5**:211-237. DOI: 10.1007/BF00615122

[44] Watanabe T, Tanaka M. Thermal plasma processing for functional nanoparticle synthesis. In: 16th ASEAN Regional Symposium on Chemical Engineering; 2009. p. 4

[45] Girshick SL, Chiu CP, McMurry PH. Modelling particle formation and growth in a plasma synthesis reactor. Plasma Chemistry and Plasma Processing. 1988;**8**:145-157. DOI: 10.1007/BF01016154

Chapter 6

Chemical Detection of Short-Lived Species Induced in Aqueous Media by Atmospheric Pressure Plasma

Yury Gorbanev and Annemie Bogaerts

Abstract

Non-thermal atmospheric pressure plasmas are widely used in biomedical research and clinical applications. Such plasmas generate a variety of reactive oxygen and nitrogen species upon interaction with ambient surroundings. These species further interact with a biological substrate and are responsible for the biomedical effects of plasma. Liquid water is an essential part of any biological systems. Some of the most reactive species induced by plasma in aqueous media are radicals and atoms. Hence, the presence of certain chemical components in a plasma 'cocktail' presents an important task for both understanding and further development of plasma systems with specific purposes. In this chapter, we discuss various methods of detection of the plasma-generated short-lived reactive species. We dissert various plasma-induced radicals and atoms (•OH, $O_2^{\bullet -}$/•OOH, •NO, O), together with non-radical short-lived species ($^{-}OONO$, O_3, $^1O_2$). Electron paramagnetic resonance (EPR) is the most direct method of radical detection in water-based media. Special attention is paid to the limitations of the detection methods, with an emphasis on spin trapping used in EPR analysis.

Keywords: plasma-liquid systems, reactive species, free radicals, spin trapping, electron paramagnetic resonance

1. Introduction

Low-temperature, or 'cold' atmospheric pressure plasmas (CAPs) are gaining increasing attention in diverse fundamental and applied scientific activities [1]. The research on the industrial applications of cold plasma includes its use in plasma-assisted catalysis and thin film deposition [1–3], wastewater treatment [4, 5], photoresist removal [6], pre-treatment of polymeric solutions for the production of enhanced nano-fibres [7], etc.

Biomedical applications are the most burgeoning field of CAP research [8]. Cold plasma is used to modify or produce surfaces with high bacterial resistance, an important property in a clinical setting [8, 9]. Other applications include various sterilisation processes [10], deactivation of bacteria and viruses [11, 12], wound healing [8], and the emerging CAPs cancer treatment [8, 13]. The effects of CAPs on biological substrates are largely defined by the reactive oxygen and nitrogen species (RONS) such as •OH, $O_2^{\bullet-}$/•OOH, O_3, $^1O_2$, O, •NO, H_2O_2, $ONOO^-$, NO_2^- [14]. These species are formed either in the plasma itself, or upon its interaction with surrounding air [14, 15]. Water is an essential component of every biological system. Thus, information on both the composition of the mixture of RONS created by plasma, and their interactions with aqueous media is extremely important for tailoring-desired plasma effects [16].

Chemical modelling coupled with various analytical techniques (optical spectroscopy methods, mass spectrometry, etc.) is used to assess the composition of the gas phase plasma [1, 14, 17]. Recent works in computational chemistry have addressed the interaction of gas phase RONS with and within aqueous media [18, 19]. However, monitoring of the reactive species in liquid is paramount for benchmarking of the models. Most importantly, it provides experimental, and hence the most direct information on RONS present in the liquid.

Two main paths are used for CAP utilisation in biomedical research and applications: first, pre-treatment of a relevant medium with further application to the biological target [20, 21] and second, direct application of plasma treatment to cells in aqueous media [22], or 'dry' cells [23] (We note that generally all cells are grown in culture medium and/or washed prior to plasma exposure; thus, the 'dry' cell surface is never devoid of water, even less, so is a tissue in clinical applications [24]). In the first scenario, the effects of plasma are attributed to long-lived chemical species, which can remain in solution after plasma treatment, such as H_2O_2, NO_2^-, and NO_3^- [20, 25]. These long-lived species are usually detected using a variety of analytical techniques, of which colorimetry is commonly employed [20, 22]. The second path implies the presence of short-lived radical and atomic species: O, •NO, •OH, $O_2^{\bullet-}$, as well as non-radical chemical compounds such as, e.g., singlet oxygen $^1O_2$. Aside from creating direct oxidative stress [14], these species can regulate various cellular processes by, e.g., altering cellular uptake of metal ions [26]. These short-lived radicals were also shown to initiate radical reactions in liquid media [27].

This chapter disserts methods of detection, identification and quantification of short-lived chemical species in solutions in contact with CAPs.

2. Detection of plasma-generated RONS in liquids

2.1 Hydroxyl and superoxide radicals

Upon interaction with water and oxygen moieties, CAPs generate hydroxyl radicals •OH and superoxide radical anions $O_2^{\bullet-}$. These short-lived species possess highly oxidising and cytotoxic properties, and are suggested to be one of the main causes of biomedical activity of cold plasma [14, 28].

In aqueous systems, these radicals are often detected using optical methods. These methods usually employ induction of colour or degradation of dyes (colorimetry) [29]. In this method, a coloured dye is degraded by plasma-generated species, and the loss of colour is quantified using UV-vis spectrophotometry. Some of the most commonly used dyes used to assess ROS are methylene blue and methylene red [29, 30], often used in research associated with CAPs for water treatment and pollutant removal [30, 31]. However, the main difficulty in employing this method is the non-specific degradation of such dyes. For example, decolouration of methylene blue not only occurs via reaction with •OH radicals but also with other CAP-induced species, e.g., ozone [32]. Superoxide radical anions can also be detected by degradation of dyes [33] with the same limitations.

Similarly, induction/decay of fluorescent properties of chemical molecules is also used. Degradation of some fluorophores was used to detect oxygen-centred radicals on surfaces in contact with processing plasmas [34].

Terephthalic acid (TA) has been reported to detect •OH radicals in plasma-liquid systems [35–37]. The method is based on the induced fluorescence due to the formation of hydroxy-substituted TA (**Scheme 1**). This presents a simpler and more selective method of detection of the hydroxyl radicals, as demonstrated by Attri and co-workers [37]. However, possible oxidation of terephthalic acid by RONS [38] is usually ignored. Another method based on aromatic hydroxylation was recently suggested by Zhang et al. Using salicylic acid as a substrate, the authors obtained mono- and disubstituted products, which were analysed by HPLC [39].

Other alternatives include different chemical probes, such as, e.g., dimethylsulphoxide, followed by quantification of the formed formaldehyde HCHO or methanesulfinic acid [35, 40] (**Scheme 2**). However, other reactions leading to both production and degradation of HCHO can occur in a plasma-liquid system, as was shown by Ma et al. [40]. The kinetics of these complex chemical processes needs to be considered for quantitative assessment of the hydroxyl radical in liquids.

Shirai et al. observed chemiluminescence at the plasma-liquid interface when alkaline solutions of luminol were exposed to CAP, presumably due to reactions with •OH and O_2•$^-$ [41]. Bekeschus et al. compared the amount of superoxide produced by different plasmas in liquid using colorimetric analysis with cytochrome C [42].

Furthermore, the information on the availability of radical species in liquids exposed to CAPs is often obtained using electron paramagnetic resonance (EPR) spectroscopy. EPR is the most direct method of radical detection in liquids [43, 44]. The method is based on the detection of paramagnetic species, e.g., free radicals with an unpaired electron. In EPR, several methods of radical detection are used [45]. Free radicals such as hydroxyl and superoxide are very short-lived and cannot be detected directly. Hence, spin probes and spin traps are employed. An example

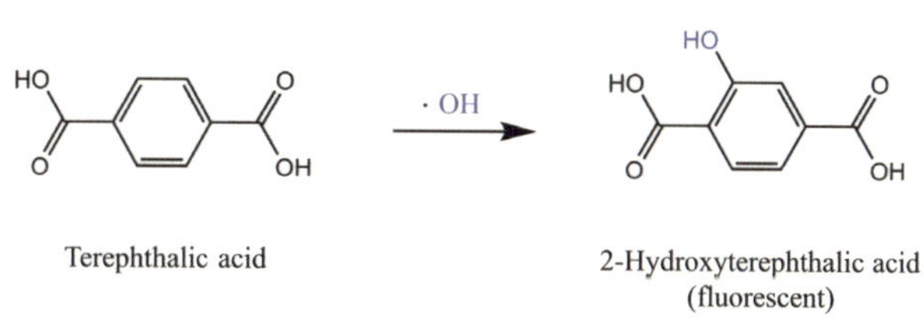

Scheme 1.
Detection of the OH radical using induced fluorescence via the reaction with terephthalic acid.

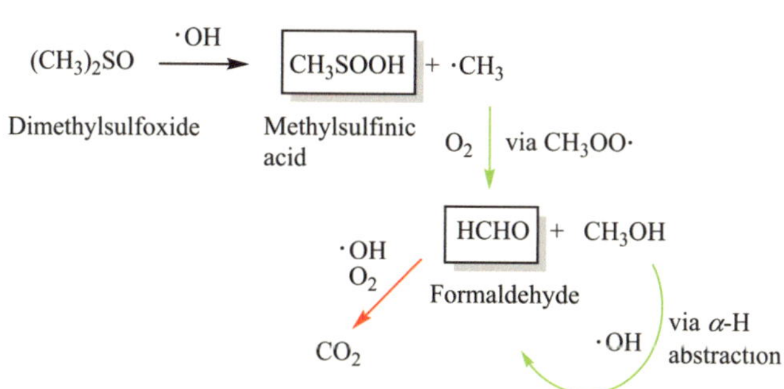

Scheme 2.
Reactions of dimethylsulphoxide used in detection of the •OH radical. Production and loss pathways of HCHO are indicated with green and red arrows, respectively.

of a spin probe is 1-hydroxy-3-methoxycarbonyl-2,2,5,5-tetramethylpyrrolidine (CMH). CMH is a cyclic hydroxylamine, which reacts with superoxide radicals to form a nitroxide radical detected by EPR: =N—OH → =N—O• [46]. However, this method is mostly used in biological systems rather than plasma-liquid systems. The selectivity is not explicitly known, and possible difficulties may arise from interferences by other plasma-induced ROS.

The second method is the formation of radical adducts in reactions of spin traps (organic molecules, usually nitrones) with free radicals (**Scheme 3**). The formed spin adducts are organic nitroxides with longer half-lives compared to the analysed radicals, and thus detectable by EPR [47, 48]. In the past decade, the spin trapping of CAP-induced radicals in liquids has gained vast attention. Numerous groups have performed detection of •OH and $O_2^{\bullet -}$/•OOH radicals in aqueous media by EPR. Tani et al., Takamatsu et al. and Uchiyama et al. performed the detection of radical species in plasma-treated liquids by using various spin traps [49–51]. Tresp et al. assessed the concentrations of the •OH and $O_2^{\bullet -}$ radicals by monitoring the radicals adducts of 5-tert-butoxycarbonyl-5-methyl-1-pyrroline *N*-oxide (BMPO) and 5,5-dimethyl-1-pyrroline *N*-oxide (DMPO) spin traps in liquid samples [52]. We have previously performed detection of hydroxyl and superoxide radicals, as well as hydrogen atoms, in CAP-treated water with DMPO, 5-(diethoxyphosphoryl)-5-methyl-1-pyrroline *N*-oxide (DEPMPO) and *N*-benzylidene-*tert*-butylamine *N*-oxide (PBN) spin traps [15, 22, 44]. We note that according to their pKa values, superoxide radical exists in its anion form $O_2^{\bullet -}$ in a physiological solution (pH 7–7.6), but the radical adducts are protonated forms —OOH [53].

The advantage of spin trapping over other methods is that the non-selectivity of a spin trap is not a drawback. Indeed, the characteristic features of the EPR signals are different for different radical adducts due to the hyperfine coupling: interaction with magnetic moments of nearby nuclei with non-zero spin numbers (^{14}N, ^{1}H, ^{13}C, ^{17}O, etc.). The hyperfine values of an adduct therefore depend on the structure of a specific adduct: DMPO-OH and DEPMPO-OH have different chemical structures and thus different features of their EPR spectra (see, e.g., [15, 44]). Also, if a spin trap (e.g., DEPMPO) forms adducts with several radicals such as •OH and $O_2^{\bullet -}$, the amounts of both adducts can be obtained from the same EPR spectrum [45]. This feature enables studying the source of the radicals produced by CAPs by using isotopically labelled water ($H_2{}^{17}O$, $^{2}H_2O$) to distinguish between the radicals formed from the gas phase water and the liquid water [15, 44, 54]. More recently, the use of PBN and DMPO spin traps helped identify the nature of the radicals generated by CAPs from organic solvents (chloroform and *N,N*-dimethylformamide) [7].

However, despite being a very versatile and the most direct method of radical detection in liquids, spin trapping and EPR analysis have limitations, which need to be considered in experimental settings. Before proceeding to the detection of other short-lived species, we first address in the following section the factors limiting the applicability of spin trapping and EPR analysis in CAP-liquid systems.

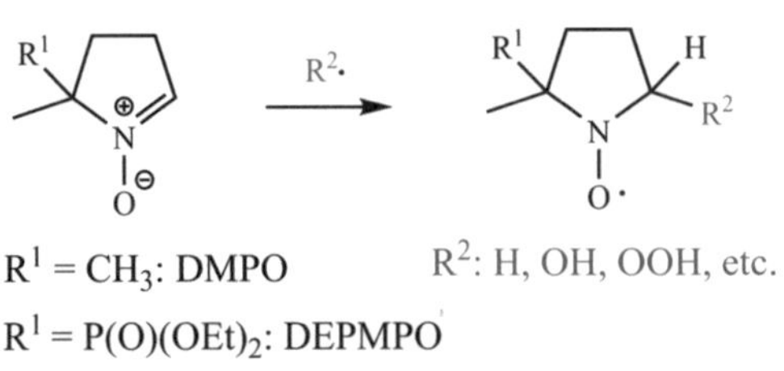

Scheme 3.
Formation of nitroxide radical adducts in reactions with CAP-induced radicals by DMPO and DEPMPO spin traps.

2.2 Limitations of spin trapping

Quantification of the radical adduct (i.e., concentration in the analysed solution) can be achieved by calibrating an EPR spectrometer with solutions of a stable radical compound, e.g., a stable nitroxide (2,2,6,6-tetramethylpiperidin-1-yl)oxyl (TEMPO) or its derivatives [15, 22, 44]. However, obtaining concentration values of radicals (rather than the formed adduct) in liquids is hindered due to various side reactions of the analysed radicals: recombination (e.g., •OH into H_2O_2), reactions with scavengers (solvated electrons, hydrogen atoms, possible physiological media components [22, 37, 42, 48]). Comparison of reaction rate coefficients of these processes needs to be performed. Otherwise, EPR with spin trapping provides only semi-quantitative data: the amounts of a radical adduct formed under different conditions follow the same trend as the initial concentration of the free radical.

Different spin traps have different affinity towards different radicals expressed by the rate coefficients of the respective reactions [44, 45]. Moreover, the stability of the formed radical adducts varies in orders of magnitude. Under physiological conditions (pH, temperature, etc.), the half-life of the radical adduct with superoxide radical of the DMPO spin trap is 45 s, with DEPMPO, it is 14 min, and with 3,5-dimethyl-5-(iso-propoxycarbonyl)-1-pyrroline *N*-oxide (3,5-DIPPO), it is 55 min [55]. The decay of spin adducts occurs naturally in liquids and depends on many factors: temperature, concentration of the adduct and other components of the solutions, etc. The half-life of the DMPO-OOH radical adduct was shown to be a function of pH by Buettner et al. [56]. The rapid decay of DMPO-OOH proceeds via formation of DMPO-OH, and in the presence of electron acceptors, it can lead to a complete disappearance of the EPR signal due to the loss of the radical moiety (**Scheme 4**). It was also reported that the stability of DMPO-OH was substantially reduced in the presence of nitrogen oxides [57].

Hence, the choice of a spin trap is a very important factor, which may affect both the quantitative and the qualitative results. Selectivity, adduct stability and last but not least commercial availability of a spin trap should be taken into account when preparing for the analysis of radical species in liquids.

The limitations discussed above are known in biological systems, with limited production and diversity of RONS. With CAPs, various atomic and radical reactive species can be simultaneously delivered to the liquid. Our previous work showed that nitroxides can decay via reactions with the same species from which they were formed [58]. Our results demonstrated that the loss of nitroxide moiety in plasma-exposed water occurred via reactions with •OH radicals, H atoms and oxygen species (atomic oxygen and/or ozone) with pseudo-first-order kinetics (**Scheme 5**). Hence, to perform even relative measurements of RONS induced in water by CAPs, a study

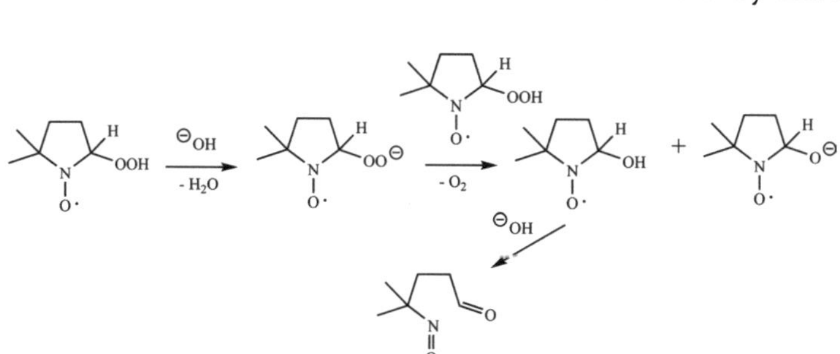

Scheme 4.
pH-dependent degradation pathway of the radical adducts of the DMPO spin trap.

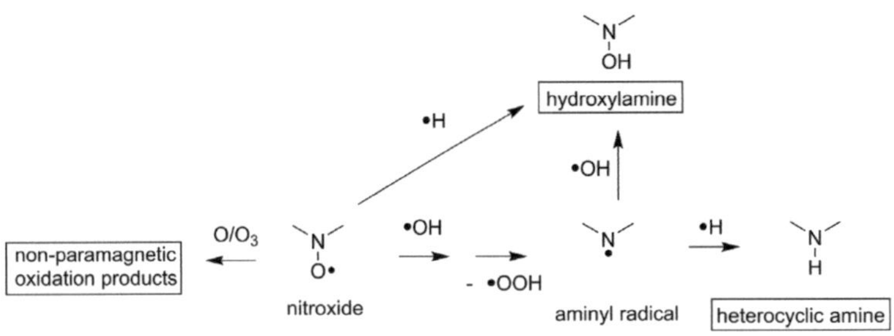

Scheme 5.
Degradation of nitroxides (radical adducts) via reactions with CAP-induced species.

of the product (spin adduct, aromatic substitution product, etc.) concentration development over time may be necessary to exclude possibilities of the increasing analyte degradation.

Another factor affecting the analysis is the potential decay of the nitrone spin traps themselves. Upon plasma exposure, it was shown that DEPMPO spin trap was partially degraded, yielding carbon-centred radicals, which were trapped by the remaining DEPMPO [42]. Similarly, PBN can undergo degradation into *tert*-butyl hydronitroxide [59].

A number of 'false-positive' results have been identified for spin trapping. For instance, a nucleophilic addition via the Forrester-Hepburn mechanism, either direct [60] or metal-catalysed [61] (**Scheme 6**), may lead to the formation of nitroxides with the same structure as the radical adducts. It is thus important to perform control experiments with no CAP-induced RONS, to assess the possible interference from such reactions.

2.3 Solvated electrons

Many types of plasma set-ups either have discharges to the surface of the liquid (e.g., floating electrode plasmas with no gas flow), or an electron-rich afterglow (most plasma jets, with the exception of COST jet-type set-ups) [15, 16, 23]. In such cases, not just the RONS, but the electrons too may interact with the CAP-exposed liquid [16, 62]. Solvated electrons contribute to the additional charge in the liquid and induce electrochemical reactions [63], affecting potential substrates.

Rumbach et al. have demonstrated that both electron transfer and neutral reactions occur when CAPs interact with aqueous media. The authors report an optical technique employing a series of individual diode lasers for the spectroscopic detection of solvated electrons [63, 64]. Another (chemical) approach describes the analysis of the pH of the saline solutions in an electrochemical system where a cathode is substituted by a plasma jet. In such system, the pH changes as a result of the chlor-alkali process initiated by plasma electrons [65]. To the best of our knowledge, no other works describe chemical detection of solvated electrons in plasma-liquid systems.

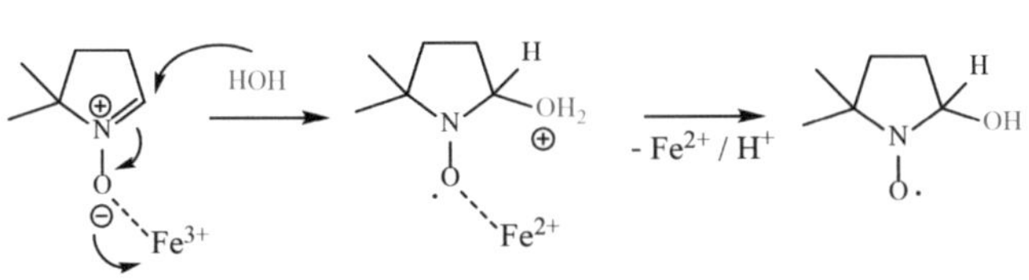

Scheme 6.
Metal-catalysed Forrester-Hepburn mechanism of nucleophilic addition of a water molecule to the DMPO nitrone spin trap, leading to the formation of the nitroxide with the DMPO-OH structure.

Several reviews and reports describe possibilities of chemical detection of solvated electrons [66, 67]. The methods include indirect detection, e.g., EPR analysis of the DMPO-spin trapped benzyl radicals, which were formed upon reaction of benzyl chloride with solvated electrons [68]. These methods could prove very useful in plasma-liquid systems, although their direct applicability (limitations due to selectivity, etc.) needs to be determined.

2.4 Atomic oxygen, singlet oxygen and ozone

Some of the most reactive and biologically relevant species created by CAPs are atomic oxygen O, ozone O_3 and singlet oxygen $^1O_2$ [14, 69]. It has been shown by Benedikt and co-workers that oxygen atoms are delivered to exposed liquid solutions from the gas phase plasma, where they are generated [70, 71], while computational results indicate that these extremely reactive atomic species can also be formed inside the liquid from hydroxyl radicals [18]. Singlet oxygen and ozone are generally considered short-lived (compared to e.g. hydrogen peroxide), although they are more stable than the atomic or radical species.

Ozone is often detected colorimetrically using the degradation of coloured dyes, e.g., methylene blue. This method of ozone detection is highly non-selective, since the dyes can be degraded by other CAP-produced RONS, including •OH radicals [32]. Kovačević et al. measured ozone delivery to the plasma-exposed liquid using iodometry, while the solubilised ozone remaining in the liquid after the CAP treatment was detected with decolourisation of indigo trisulphonate [72]. However, recently, Tarabová et al. reported that indigo dyes decay in reactions with other RONS, e.g., secondary •OH radicals produced in liquids after the plasma exposure [73]. The non-selectivity of the iodometric method is due to the other oxidising RONS [72]. Fluorescent probes have also been used to detect ozone in plasma-liquid systems, albeit not without selectivity issues [74].

Benedikt and co-workers detected oxygen atoms in aqueous solutions of phenol, with further MS analysis of the formed hydroxylated products [70, 71]. However, hydroxylation of phenol can also occur in a reaction with hydroxyl radicals [73, 75].

EPR detection of a combination of CAP-induced O, $^1O_2$ and O_3 in aqueous media was performed by Takamatsu et al. The authors used 2,2,5,5-tetramethyl-3-pyrroline-3-carboxamide (TPC) as a chemical detector of oxygenated species in liquid [50]. TPC is oxidised to produce a stable nitroxide, which can be detected by EPR. While the oxidation of TPC was non-selective, the addition of sodium azide NaN_3 as a singlet oxygen scavenger allowed to distinguish between the TPC oxidised by $^1O_2$ and by all other species.

We have previously used 2,2,6,6-tetramethylpiperidine (TEMP), another cyclic amine, to detect $O/^1O_2/O_3$ produced by an atmospheric pressure plasma via its oxidation to 2,2,6,6-tetramethylpiperidine 1-oxyl (TEMPO) [22, 44]. Adding NaN_3 let us individually assess the concentrations of $^1O_2$ and O/O_3 induced by CAPs (**Scheme 7**). Our work showed that although TEMPO was not produced in reactions with H_2O_2 or $O_2•^-$, it could be formed by other plasma-induced RONS: ozone and possibly by atomic oxygen [44].

Later, Elg et al. demonstrated that most TEMPO was in fact formed by atomic oxygen by comparing ozone densities in the gas phase with concentrations of the formed TEMPO [76]. However, the main contributor to the production of a stable nitroxide in these reactions would depend on the densities of O and O_3 (and therefore, a specific plasma set-up) in each case. Other limitations of this method are related to the loss reactions of nitroxides [58], as described above (see **Scheme 5**).

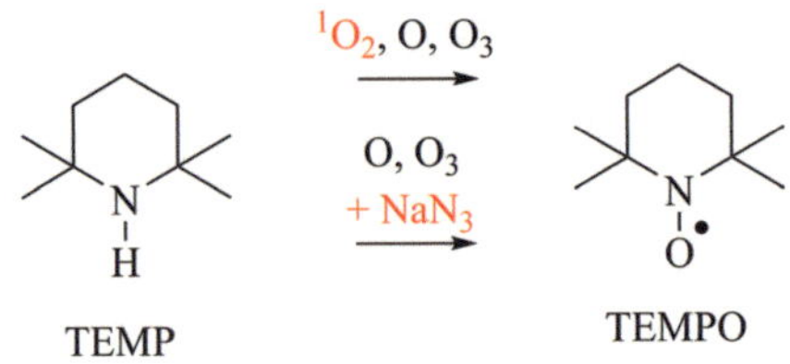

Scheme 7.
Detection of singlet oxygen, atomic oxygen and ozone by oxidation of TEMP. NaN_3 is used as a specific scavenger for $^1O_2$.

2.5 Nitrogen oxides and peroxynitrite

Nitric oxide •NO is considered one of the CAP-induced RONS responsible for bactericidal effects and wound healing. Furthermore, its reactions with and within aqueous media produce a large variety of secondary RONS [77]. Unlike the more persistent products of the •NO transformations such as NO_2^- and NO_3^-, which are detected colorimetrically [20, 48, 72, 73], the radical nitric oxide itself is monitored using spin trapping and EPR. Many EPR methods of detection of nitric oxide are known in biological milieu. Among these are the use of dithiocarbamate metal complexes, oxidation of nitronyl nitroxides, etc. [78].

In plasma-liquid systems, •NO has been detected intracellularly using fluorescent probes [51, 79] and directly in media by EPR with iron complexes of *N*-methyl-D-glucamine dithiocarbamate (MGD) [50]. In the latter reaction, chelated Fe^{2+} ions form paramagnetic complexes with •NO. However, it was shown by Tsuchiya et al. [80] that this reaction is not selective: $(MGD)_2Fe^{2+}$ complex reacts with the nitrite anion NO_2^- with the oxidation of Fe^{2+} to Fe^{3+}, eventually leading to the formation of the $(MGD)_2Fe^{2+}$ adduct with •NO (**Scheme 8**). Another limitation is the possible oxidation of the iron ion to Fe^{3+} (which forms a non-paramagnetic complex with •NO) by other plasma RONS, and thus the necessity to use large amounts of a reducing agent [58].

Another method to detect nitric oxide in CAP-liquid systems is the transformation of nitronyl nitroxides such as 2-phenyl-4,4,5,5-tetramethylimidazoline-1-oxyl 3-oxide (PTIO) or its derivatives [48, 51, 58]. PTIO, a stable nitroxide radical, reacts with nitric oxide to form 2-phenyl-4,4,5,5-tetramethylimidazoline 1-oxyl (PTI), an imino nitroxide radical (**Scheme 9**). EPR analysis with deconvolution of the spectra via radical signal simulations allows differentiating between the two radicals [48, 78]. We previously showed that the limitations of the method are related to the nitroxide decay pathways (see above). The other issue is the reverse transformation (oxidation) of PTI to PTIO in oxygen-containing plasmas, even in the absence of nitrogen [58]. This makes detection of nitric oxide in plasma-liquid systems an extremely difficult task, when both the absence of the detectable •NO (PTIO) and its presence (a 'false positive' with dithiocarbamates) can be due to the limitations of each method.

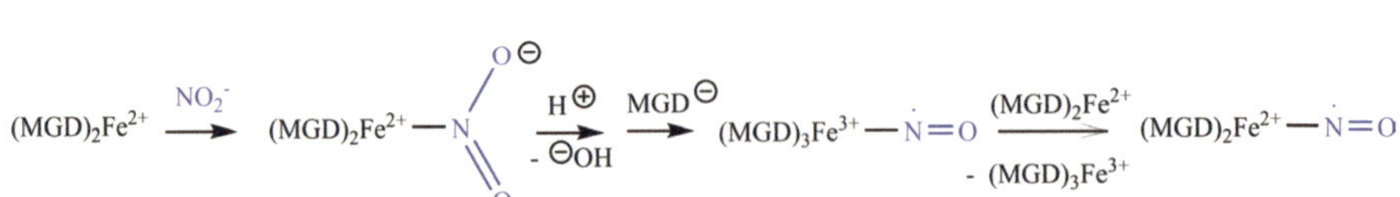

Scheme 8.
Reaction pathway of the $(MGD)_2Fe^{2+}$ complex interaction with the NO_2^- anion to yield a paramagnetic adduct $(MGD)_2Fe^{2+}$-NO.

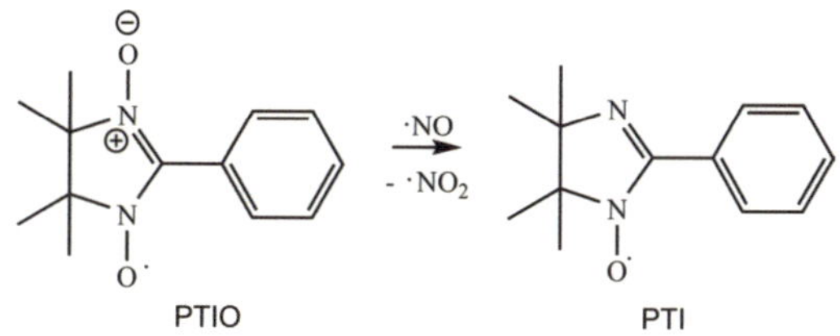

Scheme 9.
Reaction of nitronyl nitroxide PTIO with nitric oxide, yielding imino nitroxide PTI.

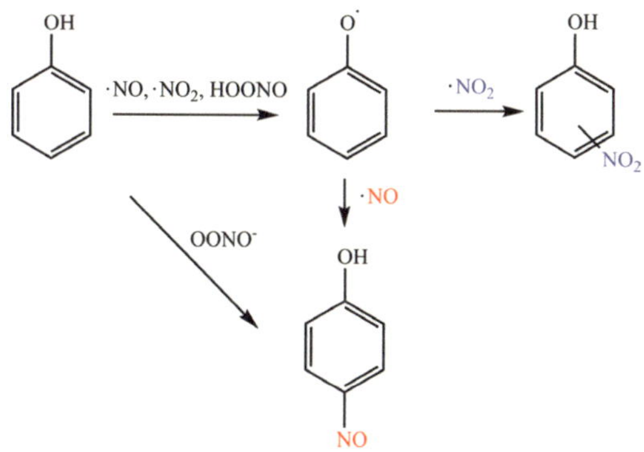

Scheme 10.
Reactions of nitration and nitrosation of phenol by nitric oxide, nitrogen dioxide and peroxynitrite.

Lukes et al. used nitrosation and nitration reactions of phenol as a technique to detect post-plasma discharge formation of nitrogen oxides (•NO and •NO_2) in aqueous media [75] (**Scheme 10**). It can be used to detect these two radicals directly during plasma exposure, or the secondary radicals formed during the decay of peroxynitrite (see below). As far as we know, no other methods of •NO_2 detection have been reported specifically in plasma-treated solution, although other ways of detecting nitrogen dioxide in aqueous media are available in literature: e.g., spin trapping with nitrones [81] or nitroalkanes [82].

The peroxynitrite anion $ONOO^-$ is another reactive species induced by CAPs in water-based media [14, 72, 83]. Lukes et al. were the first to evaluate its formation and stability in aqueous solutions exposed to plasma. Peroxynitrite decayed rapidly in acidic conditions via peroxynitrous acid decomposition into •OH and •NO_2, which were detected via phenol derivatisation [75]. The instability of peroxynitrite in neutral pH was later demonstrated by Weltmann and co-workers [84, 85]. Here, peroxynitrite was detected by exposing solutions of L-tyrosine to CAP and further MS analysis of the formed 3-nitrotyrosine. The authors have also assessed the interferences from the •NO radicals by introducing an •NO donor [85]. When L-tyrosine was added to plasma-treated solutions after the exposure, no nitration product was detected, suggesting the short-lived nature of peroxynitrite under the applied conditions. Girard et al. developed a method of direct spectrophotometrical detection of peroxynitrite based on its absorption properties in the UV region [86]. Here, peroxynitrite was only detected in highly basic solutions, confirming its short life at neutral pH. Xu et al. used 1-hydroxy-2,2,6,6-tetramethyl-4-oxo-piperidine (a hydroxylamine) to detect both superoxide radical anion and peroxynitrite anion via EPR analysis of the formed 4-oxo-2,2,6,6-tetramethylpiperidine 1-oxyl [87] (distinguishing between the two is not possible with this method [46, 87]).

2.6 Hypochlorite formed in physiological media

Both the physiological media used in research in vitro (PBS, DMEM, etc.) and the blood and tissues in vivo have high concentration of chloride salts. Chloride anions can act as scavengers of the CAPs-generated RONS in media, e.g., the •OH radicals [28, 88]. It has been suggested by different research groups that chlorinated species such as hypochlorite ClO^- (short-lived under physiological conditions [84]) can be formed in solutions upon interaction with the plasma RONS [79, 85, 89]. Wende et al. have assessed the effects of the potential ClO^- presence in situ on the overall biomedical efficacy of CAPs [85]. Recently, Piskarev et al. showed that the ClO^- anion is formed in plasma-treated saline solutions via reactions of the Cl^- anion with superoxide or hydroxyl radicals from plasma [90].

The detection of hypochlorite in solutions was performed by direct UV detection [90] or by exposure of a solution of L-tyrosine to CAPs [84, 85]. The latter method is based on the MS detection of the chlorinated product of L-tyrosine. The UV detection requires large concentrations of hypochlorite, while the MS method is highly sensitive with a low detection limit. However, Bekeshus et al. detected no hypochlorite formation with their plasma set-up [84]. The formation of hypochlorite (formed from the initial CAP-generated RONS in liquid media), therefore, depends on the particular plasma set-up and application conditions.

3. Conclusion and perspectives

A plethora of methods are available for the detection of reactive oxygen and nitrogen species induced in liquids by plasma. Chemical 'detector' systems based on colorimetry, fluorescence, (LC-)MS analysis of the products and EPR spin trapping are some of the techniques used in the detection of atomic, radical, molecular and ionic short-lived RONS. Each method has its potential and limitations; the latter associated with the decay of the products, low selectivity and other factors.

Despite several limitations, spin trapping coupled with EPR analysis as an analytical method of radical detection in CAPs systems has a very high value as the most direct method of radical detection in liquids. Aside from the liquid media itself, it has a potential to be used in plasma-gel systems, which mimic interaction of CAPs with tissue [91], if gels are formed from nitrone molecules [92]. The availability of EPR equipment is not necessarily crucial to perform spin trapping of RONS. An interesting alternative to EPR analysis of radical adducts was demonstrated by Guo et al. and Tuccio et al., who used liquid chromatography and mass spectrometry systems to analyse the adducts of oxygen-centred radicals of the DMPO and DEPMPO spin traps [93, 94].

The emerging role of the chlorinated species in CAP-treated media is gaining attention. ClO^- anion, a biomedically relevant species, has been monitored using UV absorption spectroscopy and modification of tyrosine. However, other detection methods may need to be used, especially in cases when hypochlorite is not detected. For example, colorimetric analysis on the oxidation of 3,3′,5,5′-tetramethylbenzidine is a simple technique [95], albeit with yet unknown limitations due to the presence of other oxidising RONS. Moreover, the highly oxidative nature of the hypochlorite anion in plasma-treated physiological solutions is often emphasised, but the possibility of formation of other anions such as ClO_2^-, ClO_3^- and ClO_4^- has not been addressed. Since these species are cytotoxic, monitoring them in physiological media exposed to plasma can provide valuable information on the biomedical effects of CAPs. Possible analyses can include ion chromatography [96, 97].

Finally, some other short-lived yet highly reactive species are overlooked in the current research. Among these is the carbonate radical anion CO_3•.: a very potent oxidising agent causing DNA damage [98], which can be formed in a reaction of peroxynitrite with ambient CO_2 [99, 100]. The determination of this and other reactive species will aid in completing the picture of the plasma-produced 'cocktail' of reactive species. It can facilitate both the understanding of the existing CAP devices and their effects, and the development of new plasma systems with dedicated RONS concentrations for specific applications.

Acknowledgements

Yury Gorbanev thanks the European Marie Sklodowska-Curie Individual Fellowship within Horizon2020 ('LTPAM', grant no. 743151) for supporting his work.

Conflict of interest

The authors have no conflict of interest to declare.

Author details

Yury Gorbanev* and Annemie Bogaerts
Research Group PLASMANT, Department of Chemistry, University of Antwerp, Antwerpen, Belgium

*Address all correspondence to: yury.gorbanev@uantwerpen.be

References

[1] Adamovich I, Baalrud SD, Bogaerts A, Bruggeman PJ, Cappelli M, Colombo V, Czarnetzki U, Ebert U, Eden JG, Favia P, Graves DB, Hamaguchi S, Hieftje G, Hori M, Kaganovich ID, Kortshagen U, Kushner MJ, Mason NJ, Mazouffre S, Mededovic Thagard S, Metelmann HR, Mizuno A, Moreau E, Murphy AB, Niemira BA, Oehrlein GS, Petrovic ZL, Pitchford LC, Pu YK, Rauf S, Sakai O, Samukawa S, Starikovskaia S, Tennyson J, Terashima K, Turner MM, Van De Sanden MCM, Vardelle A. The 2017 plasma roadmap: Low temperature plasma science and technology. Journal of Physics D: Applied Physics. 2017;**50**:323001. DOI: 10.1088/1361-6463/aa76f5

[2] Bogaerts A, Neyts EC. Plasma technology: An emerging Technology for Energy Storage. ACS Energy Letters. 2018;**3**:1013-1027. DOI: 10.1021/acsenergylett.8b00184

[3] Kakiuchi H. Atmospheric-pressure low-temperature plasma processes for thin film deposition. Journal of Vacuum Science & Technology, A: Vacuum, Surfaces, and Films. 2014;**32**:030801. DOI: 10.1116/1.4828369

[4] Vanraes P, Nikiforov AY, Leys C. Electrical discharge in water treatment technology for micropollutant decomposition. In: Mieno T, editor. Plasma Science and Technology. London: IntechOpen; 2016. pp. 429-478. DOI: 10.5772/61830

[5] Vanraes P, Wardenier N, Surmont P, Lynen F, Nikiforov A, Van Hulle SWH, Leys C, Bogaerts A. Removal of alachlor, diuron and isoproturon in water in a falling film dielectric barrier discharge (DBD) reactor combined with adsorption on activated carbon textile: Reaction mechanisms and oxidation by-products. Journal of Hazardous Materials. 2018;**354**:180-190. DOI: 10.1016/j.jhazmat.2018.05.007

[6] West A, van der Schans M, Xu C, Cooke M, Wagenaars E. Plasma Sources Science and Technology. 2016;**25**:02LT01. DOI: 10.1088/0963-0252/25/2/02LT01

[7] Rezaei F, Gorbanev Y, Chys M, Nikiforov A, Van Hulle SWH, Cos P, Bogaerts A, De Geyter N. Investigation of plasma-induced chemistry in organic solutions for enhanced electrospun PLA nanofibers. Plasma Processes and Polymers. 2018;**15**:e1700226. DOI: 10.1002/ppap.201700226

[8] Weltmann K-D, von Woedtke T. Plasma medicine—Current state of research and medical application. Plasma Physics and Controlled Fusion. 2017;**59**:014031. DOI: 10.1088/0741-3335/59/1/014031

[9] Bazaka K, Jacob MV, Chrzanowski W, Ostrikov K. Anti-bacterial surfaces: Natural agents, mechanisms of action, and plasma surface modification. RSC Advances. 2015;**5**:48739-48759. DOI: 10.1039/C4RA17244B

[10] Shintani H, Sakudo A, editors. Gas Plasma Sterilization in Microbiology: Theory, Applications, Pitfalls and New Perspectives. Vol. 156. Poole: Caister Academic Press; 2016. DOI: 10.21775/9781910190258

[11] Zhang Z, Xu Z, Cheng C, Wei J, Lan Y, Ni G, Sun Q, Qian S, Zhang H, Xia W, Shen J, Meng Y, Chu PK. Bactericidal effects of plasma induced reactive species in dielectric barrier gas–liquid discharge. Plasma Chemistry and Plasma Processing. 2017;**37**:415-431. DOI: 10.1007/s11090-017-9784-z

[12] Aboubakr HA, Gangal U, Youssef MM, Goyal SM, Bruggeman PJ. Inactivation of virus in solution by cold atmospheric pressure plasma: Identification of chemical inactivation pathways. Journal of Physics D:

Applied Physics. 2016;**49**:204001. DOI: 10.1088/0022-3727/49/20/204001

[13] Yan D, Sherman JH, Keidar M. Cold atmospheric plasma, a novel promising anti-cancer treatment modality. Oncotarget. 2017;**8**:15977-15995. DOI: 10.18632/oncotarget.13304

[14] Lu X, Naidis GV, Laroussi M, Reuter S, Graves DB, Ostrikov K. Reactive species in non-equilibrium atmospheric-pressure plasmas: Generation, transport, and biological effects. Physics Reports. 2016;**630**:1-84. DOI: 10.1016/j.physrep.2016.03.003

[15] Gorbanev Y, Verlackt CCW, Tinck S, Tuenter E, Foubert K, Cos P, Bogaerts A. Combining experimental and modelling approaches to study the sources of reactive species induced in water by the COST RF plasma jet. Physical Chemistry Chemical Physics. 2018;**2018**(20):2797-2808. DOI: 10.1039/C7CP07616A

[16] Bruggeman PJ, Kushner MJ, Locke BR, Gardeniers JGE, Graham WG, Graves DB, Hofman-Caris RCHM, Maric D, Reid JP, Ceriani E, Fernandez Rivas D, Foster JE, Garrick SC, Gorbanev Y, Hamaguchi S, Iza F, Jablonowski H, Klimova E, Kolb J, Krcma F, Lukes P, Machala Z, Marinov I, Mariotti D, Mededovic Thagard S, Minakata D, Neyts E, Pawlat J, Petrovic ZL, Pflieger R, Reuter S, Schram DC, Schröter S, Shiraiwa M, Tarabova B, Tsai PA, Verlet JRR, von Woedtke T, Wilson KR, Yasui K, Zvereva G. Plasma-liquid interactions: A review and roadmap. Plasma Sources Science and Technology. 2016. DOI: 25:053002. DOI: 10.1088/0963-0252/25/5/053002

[17] Alves LL, Bogaerts A, Guerra V, Turner MM. Foundations of modelling of nonequilibrium low-temperature plasmas. Plasma Sources Science and Technology. 2018;**27**:023002. DOI: 10.1088/1361-6595/aaa86d

[18] Verlackt CCW, Neyts EC, Bogaerts A. Atomic scale behavior of oxygen-based radicals in water. Journal of Physics D: Applied Physics. 2017;**50**:11LT01. DOI: 10.1088/1361-6463/aa5c60

[19] Verlackt CCW, Van Boxem W, Bogaerts A. Transport and accumulation of plasma generated species in aqueous solution. Physical Chemistry Chemical Physics. 2018;**20**:6845-6859. DOI: 10.1039/c7cp07593f

[20] Van Boxem W, Van der Paal J, Gorbanev Y, Vanuytsel S, Smits E, Dewilde S, Bogaerts A. Anti-cancer capacity of plasma-treated PBS: Effect of chemical composition on cancer cell cytotoxicity. Scientific Reports. 2017;**7**:16478. DOI: 10.1038/s41598-017-16758-8

[21] Yan D, Nourmohammadi N, Bian K, Murad F, Sherman JH, Keidar M. Stabilizing the cold plasma-stimulated medium by regulating medium's composition. Scientific Reports. 2016;**6**:26016. DOI: 10.1038/srep26016

[22] Privat-Maldonado A, Gorbanev Y, O'Connell D, Vann R, Chechik V, van der Woude MW. Non-target biomolecules Alter macromolecular changes induced by bactericidal low-temperature plasma. IEEE Transactions on Radiation and Plasma Medical Sciences. 2018;**2**:121-128. DOI: 10.1109/TRPMS.2017.2761405

[23] Lin A, Truong B, Patel S, Kaushik N, Choi EH, Fridman G, Fridman A, Miller V. Nanosecond-pulsed DBD plasma-generated reactive oxygen species trigger immunogenic cell death in A549 lung carcinoma cells through intracellular oxidative stress. International Journal of Molecular Sciences. 2017;**18**:966. DOI: 10.3390/ijms18050966

[24] Norberg SA, Tian W, Johnsen E, Kushner MJ. Atmospheric pressure

plasma jets interacting with liquid covered tissue: Touching and not-touching the liquid. Journal of Physics D: Applied Physics. 2014;**47**:475203. DOI: 10.1088/0022-3727/47/47/475203

[25] Laroussi M. Plasma medicine: A brief introduction. Plasma. 2018;**1**:5. DOI: 10.3390/plasma1010005

[26] Sasaki S, Kanzaki M, Kaneko T. Calcium influx through TRP channels induced by short-lived reactive species in plasma-irradiated solution. Scientific Reports. 2016;**6**:25728. DOI: 10.1038/srep25728

[27] Gorbanev Y, Leifert D, Studer A, O'Connell D, Chechik V. Initiating radical reactions with non-thermal plasmas. Chemical Communications. 2017;**53**:3685-3688. DOI: 10.1039/C7CC01157A

[28] Xu D, Liu D, Wang B, Chen C, Chen Z, Li D, Yang Y, Chen H, Kong MG. In situ OH generation from O_2^- and H_2O_2 plays a critical role in plasma-induced cell death. PLoS One. 2015;**10**:e0128205. DOI: 10.1371/journal.pone.0128205

[29] Takemura Y, Yamaguchi N, Hara T. Decomposition of methylene blue by using an atmospheric plasma jet with Ar, N_2, O_2, or air. Japanese Journal of Applied Physics. 2013;**52**:056102. DOI: 10.7567/JJAP.52.056102

[30] Attri P, Yusupov M, Park LLP, Koduru JR, Shiratani M, Choi EH, Bogaerts A. Mechanism and comparison of needle-type non-thermal direct and indirect atmospheric pressure plasma jets on the degradation of dyes. Scientific Reports. 2016;**6**:34419. DOI: 10.1038/srep34419

[31] Garcia MC, Mora M, Esquivel D, Foster JE, Rodero A, Jimenez-Sanchidrian C, Romero-Salguero FJ. Microwave atmospheric pressure plasma jets for wastewater treatment: Degradation of methylene blue as a model dye. Chemosphere. 2017;**180**:239-246. DOI: 10.1016/j.chemosphere.2017.03.126

[32] Zhao YY, Wang T, MacGregor SJ, Wilson MP, Given MJ, Timoshkin IV. Investigation of plasma-induced methylene blue degradation using dielectric barrier discharge. In: Proceedings of the 20th International Conference on Gas Discharges and their Applications (GD2014). 6-11 July 2014; Orleans, France. pp. 566-569

[33] Hayyan M, Hashim MA, AlNashef IM. Superoxide ion: Generation and chemical implications. Chemical Reviews. 2016;**116**:3029-3085. DOI: 10.1021/acs.chemrev.5b00407

[34] Choudhury FA. Fluorophore-based sensor for oxygen radicals in processing plasmas. Journal of Vacuum Science & Technology, A: Vacuum, Surfaces, and Films. 2015;**33**:061305. DOI: 10.1116/1.4930315

[35] Sahni M, Locke BR. Quantification of hydroxyl radicals produced in aqueous phase pulsed electrical discharge reactors. Industrial and Engineering Chemistry Research. 2006;**45**:5819-5825. DOI: 10.1021/ie0601504

[36] Kanazawa S, Kawano H, Watanabe S, Furuki T, Akamine S, Ichiki R, Ohkubo T, Kocik M, Mizeraczyk J. Observation of OH radicals produced by pulsed discharges on the surface of a liquid. Plasma Sources Science and Technology. 2011;**20**:034010. DOI: 10.1088/0963-0252/20/3/034010

[37] Attri P, Kim YH, Park DH, Park JH, Hong YJ, Uhm HS, Kim K-N, Fridman A, Choi EH. Generation mechanism of hydroxyl radical species and its lifetime prediction during the plasma-initiated ultraviolet (UV) photolysis. Scientific Reports. 2015;**5**:9332. DOI: 10.1038/srep09332

[38] Thiruvenkatachari R, Kwon TO, Jun JC, Balaji S, Matheswaran M, Moon IS. Application of several advanced oxidation processes for the destruction of terephthalic acid (TPA). Journal of Hazardous Materials. 2007;**142**:308-314. DOI: 10.1016/j.jhazmat.2006.08.023

[39] Zhang X, Zhou R, Bazaka K, Liu Y, Zhou R, Chen G, Chen Z, Liu Q, Yang S, Ostrikov K. Quantification of plasma produced OH radical density for water sterilization. Plasma Processes and Polymers. 2018;**15**:e1700241. DOI: 10.1002/ppap.201700241

[40] Ma Y, Gong X, He B, Li X, Cao D, Li J, Xiong Q, Chen Q, Chen BH, Liu QH. On the quantification of the dissolved hydroxyl radicals in the plasma-liquid system using the molecular probe method. Journal of Physics D: Applied Physics. 2018;**51**:155205. DOI: 10.1088/1361-6463/aab379

[41] Shirai N, Matsuda Y, Sasaki K. Visualization of short-lived reactive species in liquid in contact with atmospheric-pressure plasma by chemiluminescence of luminol. Applied Physics Express. 2018;**11**:026201. DOI: 10.7567/apex.11.026201

[42] Bekeschus S, Schmidt A, Niessner F, Gerling T, Weltmann K-D, Wende K. Basic research in plasma medicine - a throughput approach from liquids to cells. Journal of Visualized Experiments. 2017;**129**:e56331. DOI: 10.3791/56331

[43] Kohno M. Applications of electron spin resonance spectrometry for reactive oxygen species and reactive nitrogen species research. Journal of Clinical Biochemistry and Nutrition. 2010;**47**:1-11. DOI: 10.3164/jcbn.10-13R

[44] Gorbanev Y, O'Connell D, Chechik V. Non-thermal plasma in contact with water: The origin of species. Chemistry - A European Journal. 2016;**22**:3496-3505. DOI: 10.1002/chem.201503771

[45] Chechik V, Carter E, Murphy D. Electron Paramagnetic Resonance. Oxford, UK. Oxford University Press; 2016. p. 128 ISBN: 9780198727606

[46] Torfs E, Vajs J, Bidart de Macedo M, Cools F, Vanhoutte B, Gorbanev Y, Bogaerts A, Verschaeve L, Caljon G, Maes L, Delputte P, Cos P, Košmrlj J, Cappoen D. Synthesis and in vitro investigation of halogenated 1,3-bis (4-nitrophenyl)triazenide salts as antitubercular compounds. Chemical Biology and Drug Design. 2018;**91**: 631-640. DOI: 10.1111/cbdd.13087

[47] Pan J, Sun P, Tian Y, Zhou H, Wu H, Bai N, Liu F, Zhu W, Zhang J, Becker KH, Fang J. A novel method of tooth whitening using cold plasma microjet driven by direct current in atmospheric-pressure air. IEEE Transactions on Plasma Science. 2010;**38**:3143-3151. DOI: 10.1109/TPS.2010.2066291

[48] Chauvin J, Judée F, Yousfi M, Vicendo P, Merbahi N. Analysis of reactive oxygen and nitrogen species generated in three liquid media by low temperature helium plasma jet. Scientific Reports. 2017;7:4562. DOI: 10.1038/s41598-017-04650-4

[49] Tani A, Ono Y, Fukui S, Ikawa S, Kitano K. Free radicals induced in aqueous solution by non-contact atmospheric-pressure cold plasma. Applied Physics Letters. 2012;**100**:254103-254105. DOI: 10.1063/1.4729889

[50] Takamatsu T, Uehara K, Sasaki Y, Miyahara H, Matsumura Y, Iwasawa A, Ito N, Azuma T, Kohno M, Okino A. Investigation of reactive species using various gas plasmas. RSC Advances. 2014;**4**:39901-39905. DOI: 10.1039/C4RA05936K

[51] Uchiyama H, Zhao Q-L, Ali Hassan M, Andocs G, Nojima N, Takeda K, Ishikawa K, Hori M, Kondo T. EPR-spin trapping and flow Cytometric

studies of free radicals generated using cold atmospheric argon plasma and X-ray irradiation in aqueous solutions and intracellular milieu. PLoS One. 2015;**10**:e0136956. DOI: 10.1371/journal.pone.0136956

[52] Tresp H, Hammer MU, Winter J, Wetlmann K-D, Reuter S. Quantitative detection of plasma-generated radicals in liquids by electron paramagnetic resonance spectroscopy. Journal of Physics D: Applied Physics. 2013;**46**:435401-435408. DOI: 10.1088/0022-3727/46/43/435401

[53] Burgett RA, Bao X, Villamena FA. Superoxide radical anion adduct of 5,5-Dimethyl-1-pyrroline N-oxide (DMPO). 3. Effect of mildly acidic pH on the thermodynamics and kinetics of adduct formation. The Journal of Physical Chemistry A. 2008;**112**:2447-2455. DOI: 10.1021/jp7107158

[54] Gorbanev Y, Soriano R, O'Connell D, Chechik V. An atmospheric pressure plasma setup to investigate the reactive species formation. Journal of Visualized Experiments. 2016;**117**:e54765. DOI: 10.3791/54765

[55] Stolze K, Rohr-Udilova N, Rosenau T, Stadtmüller R, Nohl H. Very stable superoxide radical adducts of 5-ethoxycarbonyl- 3,5-dimethyl-pyrroline N-oxide (3,5-EDPO) and its derivatives. Biochemical Pharmacology. 2005;**69**:1351-1361. DOI: 10.1016/j.bcp.2005.01.019

[56] Buettner GR, Oberley LW. Considerations in the spin trapping of superoxide and hydroxyl radical in aqueous systems using 5,5-dimethyl-1-pyrroline-l-oxide. Biochemical and Biophysical Research Communications. 1978;**83**:69-74. DOI: 10.1016/0006-291X(78)90398-4

[57] Reszka KJ, McCormick ML, Buettner GR, Hart CM, Britigan BE. Nitric oxide decreases the stability of DMPO spin adducts. Nitric Oxide. 2006;**15**:133-141. DOI: 10.1016/j.niox.2006.03.004

[58] Gorbanev Y, Stehling N, O'Connell D, Chechik V. Reactions of nitroxide radicals in aqueous solutions exposed to non-thermal plasma: Limitations of spin trapping of the plasma induced species. Plasma Sources Science and Technology. 2016;**25**:055017. DOI: 10.1088/0963-0252/25/5/055017

[59] Moreno SNJ, Stolze J, Janzen EG, Mason RP. Oxidation of cyanide to the cyanyl radical by peroxidase/H_2O_2 systems as determined by spin trapping. Archives of Biochemistry and Biophysics. 1988;**265**:267-271. DOI: 10.1016/0003-9861(88)90127-0

[60] Forrester AR, Hepburn SP. Spin traps. A cautionary note. Journal of the Chemical Society C: Organic. 1971;**0**:701-703. DOI: 10.1039/J39710000701

[61] Rhodes CJ, editor. Toxicology of the Human Environment: The Critical Role of Free Radicals. Vol. 512. Boca Raton, Florida: CRC Press; 2000. ISBN: 9780748409167

[62] Richmonds C, Witzke M, Bartling B, Lee SW, Wainright J, Liu C-C, Sankaran RM. Electron-transfer reactions at the plasma-liquid interface. Journal of the American Chemical Society. 2011;**133**:17582-17585. DOI: 10.1021/ja207547b

[63] Rumbach P, Bartels DM, Sankaran RM, Go DB. The effect of air on solvated electron chemistry at a plasma/liquid interface. Journal of Physics D: Applied Physics. 2015;**48**:424001. DOI: 10.1088/0022-3727/48/42/424001

[64] Rumbach P, Witzke M, Sankaran RM, Go DB. Decoupling interfacial reactions between plasmas and liquids: Charge transfer vs plasma neutral reactions. Journal of the American

Chemical Society. 2013;**135**:16264-16267. DOI: /10.1021/ja407149y

[65] Rumbach P, Bartels DM, Sankaran RM, Go DB. The solvation of electrons by an atmospheric-pressure plasma. Nature Communications. 2015;**6**:7248. DOI: 10.1038/ncomms8248

[66] Murai H. Spin-chemical approach to photochemistry: Reaction control by spin quantum operation. Journal of Photochemistry and Photobiology, C: Photochemistry Reviews. 2003;**3**:183-201. DOI: 10.1016/S1389-5567(02)00038-2

[67] Fessenden RW. ESR of reacting radicals. Applied Magnetic Resonance. 2014;**45**:483-503. DOI: 10.1007/s00723-014-0535-5

[68] Šima J, Brezova V. Mechanism of photoinduced processes in solutions of iodo iron (III) complexes containing Schiff base ligands. Monatshefte für Chemie/Chemical Monthly. 2001;**132**:1493-1500. DOI: 10.1007/s007060170005

[69] Bekeschus S, Wende K, Hefny MM, Rödder K, Jablonowski H, Schmidt A, von Woedtke T, Weltmann K-D, Benedikt J. Oxygen atoms are critical in rendering THP-1 leukaemia cells susceptible to cold physical plasma-induced apoptosis. Scientific Reports. 2017;**7**:2791. DOI: 10.1038/s41598-017-03131-y

[70] Hefny MM, Pattyn C, Lukes P, Benedikt J. Atmospheric plasma generates oxygen atoms as oxidizing species in aqueous solutions. Journal of Physics D: Applied Physics. 2016;**49**:404002. DOI: 10.1088/0022-3727/49/40/404002

[71] Benedikt J, Hefny MM, Shaw A, Buckley BR, Iza F, Schakermann S, Bandow JE. The fate of plasma-generated oxygen atoms in aqueous solutions: Non-equilibrium atmospheric pressure plasmas as an efficient source of atomic O(aq). Physical Chemistry Chemical Physics. 2018;**20**:12037-12042. DOI: 10.1039/c8cp00197a

[72] Kovačević VV, Dojčinović BP, Jović M, Roglić GM, Obradović BM, Kuraica MM. Measurement of reactive species generated by dielectric barrier discharge in direct contact with water in different atmospheres. Journal of Physics D: Applied Physics

[73] Tarabová B, Lukeš P, Janda M, Hensel K, Šikurová L, Machala Z. Specificity of detection methods of nitrites and ozone in aqueous solutions activated by air plasma. Plasma Processes and Polymers. DOI: 10.1002/ppap.201800030

[74] Castelló Beltrán C. Fluorescent probes for selective detection of ozone in plasma applications [thesis]. Loughborough: Loughborough University; 2015

[75] Lukes P, Dolezalova E, Sisrova I, Clupek M. Aqueous-phase chemistry and bactericidal effects from an air discharge plasma in contact with water: Evidence for the formation of peroxynitrite through a pseudo-second-order post-discharge reaction of H_2O_2 and HNO_2. Plasma Sources Science and Technology. 2014;**23**:015019. DOI: 10.1088/0963-0252/23/1/015019

[76] Elg DT, Yang I-W, Graves DB. Production of TEMPO by O atoms in atmospheric pressure non-thermal plasma–liquid interactions. Journal of Physics D: Applied Physics. 2107;**50**:475201. DOI: 10.1088/1361-6463/aa8f8c

[77] Suschek CV, Opländer C. The application of cold atmospheric plasma in medicine: The potential role of nitric oxide in plasma-induced effects. Clinical Plasma Medicine. 2016;**4**:1-8. DOI: 10.1016/j.cpme.2016.05.001

[78] Hogg NN. Detection of nitric oxide by electron paramagnetic resonance spectroscopy. Free Radical Biology & Medicine. 2010;**49**:122-129. DOI: 10.1016/j.freeradbiomed.2010.03.009

[79] Li Y, Kang MH, Uhm HS, Lee GJ, Choi EH, Han I. Effects of atmospheric-pressure non-thermal bio-compatible plasma and plasma activated nitric oxide water on cervical cancer cells. Scientific Reports. 2017;**7**:45781. DOI: 10.1038/srep45781

[80] Tsuchiya K, Yoshizumi M, Houchi H, Mason RP. Nitric oxide-forming reaction between the iron-*N*-methyl-D-glucamine dithiocarbamate complex and nitrite. The Journal of Biological Chemistry. 2000;**275**:1551-1556. DOI: 10.1074/jbc.275.3.1551

[81] Nash KM, Rockenbauer A, Villamena FA. Reactive nitrogen species reactivities with nitrones: Theoretical and experimental studies. Chemical Research in Toxicology. 2012;**25**:1581-1597. DOI: 10.1021/tx200526y

[82] Reszka KJ, Bilski P, Chignell CF. Spin trapping of nitric oxide by aci anions of nitroalkanes. Nitric Oxide. 2004;**10**:53-59. DOI: 10.1016/j.niox.2004.03.001

[83] Girard P-M, Arbabian A, Fleury M, Bauville G, Puech V, Dutreix M, Santos Sousa J. Synergistic effect of H_2O_2 and NO_2 in cell death induced by cold atmospheric he plasma. Scientific Reports. 2016;**6**:29098. DOI: 10.1038/srep29098

[84] Bekeschus S, Kolata J, Winterbourn C, Kramer A, Turner R, Weltmann K-D, Bröker B, Masur K. Hydrogen peroxide: A central player in physical plasma-induced oxidative stress in human blood cells. Free Radical Research. 2014;**48**:542-549. DOI: 10.3109/10715762.2014.892937

[85] Wende K, Williams P, Dalluge J, Gaens WV, Aboubakr H, Bischof J, von Woedtke T, Goyal SM, Weltmann K-D, Bogaerts A, Masur K, Bruggeman PJ. Identification of the biologically active liquid chemistry induced by a nonthermal atmospheric pressure plasma jet. Biointerphases. 2015;**10**:029518. DOI: 10.1116/1.4919710

[86] Girard F, Badets V, Blanc S, Gazeli K, Marlin L, Authier L, Svarnas P, Sojic N, Clément F, Arbault S. Formation of reactive nitrogen species including peroxynitrite in physiological buffer exposed to cold atmospheric plasma. RSC Advances. 2016;**6**:78457-78467. DOI: 10.1039/C6RA12791F

[87] Xu H, Chen C, Liu D, Xu D, Liu Z, Wang X, Kong MG. Contrasting characteristics of aqueous reactive species induced by cross-field and linear-field plasma jets. Journal of Physics D: Applied Physics. 2017;**50**:245201. DOI: 10.1088/1361-6463/aa7118

[88] Liao CH, Kang SF, Wu FA. Hydroxyl radical scavenging role of chloride and bicarbonate ions in the H_2O_2/UV process. Chemosphere. 2001;**44**:1193-1200. DOI: 10.1016/S0045-6535(00)00278-2

[89] Hänsch MAC, Mann M, Weltmann K-D, von Woedtke T. Analysis of antibacterial efficacy of plasma-treated sodium chloride solutions. Journal of Physics D: Applied Physics. 2015;**48**:454001. DOI: 10.1088/0022-3727/48/45/45400

[90] Piskarev IM. Effects of cold plasma and UV-C radiation on isotonic solution. High Energy Chemistry. 2017;**51**:297-301. DOI: 10.1134/S0018143917040129

[91] Szili EJ, Bradley JW, Short RD. A 'tissue model' to study the plasma delivery of reactive oxygen species. Journal of Physics D: Applied

Physics. 2014;**47**:152002. DOI: 10.1088/0022-3727/47/15/152002

[92] Richards JE, Philp D. A reactive nitrone-based organogel that self-assembles from its constituents in chloroform. Chemical Communications. 2016;**52**:4995-4998. DOI: 10.1039/c6cc01259k

[93] Guo Q, Qian SY, Mason RP. Separation and identification of DMPO adducts of oxygen-centered radicals formed from organic hydroperoxides by HPLC-ESR, ESI-MS and MS/MS. Journal of the American Society for Mass Spectrometry. 2003;**14**:862-871. DOI: 10.1016/S1044-0305(03)00336-2

[94] Tuccio B, Lauricella R, Charles L. Characterisation of free radical spin adducts of the cyclic β-phosphorylated nitrone DEPMPO using tandem mass spectrometry. International Journal of Mass Spectrometry. 2006;(1):47-53. DOI: 10.1016/j.ijms.2006.02.009

[95] Guo Y, Ma Q, Cao F, Zhao Q, Ji X. Colorimetric detection of hypochlorite in tap water based on the oxidation of 3,3′,5,5′-tetramethyl benzidine. Analytical Methods. 2015;**7**:4055-4058. DOI: 10.1039/C5AY00735F

[96] Stahl R. Ion chromatographic determination of chloride, chlorate, and perchlorate in sulfuric acid solutions. Chromatographia. 1993;**37**:300-302. DOI: 10.1007/BF02278638

[97] Seiler MA, Neist U, Deister UK, Schmitz F. Trace perchlorate determination by ion chromatography. The Column. 2016;**12**:21-25

[98] Shafirovich V, Dourandin A, Huang W, Geacintov NE. The carbonate radical is a site-selective oxidizing agent of guanine in double-stranded oligonucleotides. The Journal of Biological Chemistry. 2001;**276**:24621-24626. DOI: 10.1074/jbc.M101131200

[99] Bonini MG, Radi R, Ferrer-Sueta G, Ferreira AMDC, Augusto O. Direct EPR detection of the carbonate radical anion produced from peroxynitrite and carbon dioxide. The Journal of Biological Chemistry. 1999;**274**:10802-10806. DOI: 10.1074/jbc.274.16.10802

[100] Medinas DB, Cerchiaro G, Trindade DF, Augusto O. The carbonate radical and related oxidants derived from bicarbonate buffer. IUBMB Life. 2007;**59**:255-262. DOI: 10.1080/15216540701230511